Indicators of Social and Economic Change and Their Applications

Division for Socio-economic Analysis
Sector of Social Sciences and Their Applications

unesco

ISBN 92-3-101493-5
French edition 92-3-201493-9

Published in 1977 by Unesco
7, Place de Fontenoy, 75700 Paris, France

Composed and printed in
the workshops of Unesco
Printed in France

Preface

This collection of papers should be viewed within the context of a project of the Unesco Socio-economic Analysis Division on the development of indicators of social and economic change and their use in development planning.

Unesco's programme of work on human resources indicators began in 1967 with a first meeting of experts in Warsaw later that year and the establishment of a Panel of Experts to guide the project through 1972. Many occasional papers were produced for meetings of the Panel, and a book "Toward a system of human resources indicators for less developed countries" edited by Zygmunt Gostkowski, published for the Polish Academy of Science by Ossolineum, Warsaw, which contains some of the more important studies during that period, was published in 1974. The new project on socio-economic indicators and their use in development planning marks a shift in emphasis. This shift is symptomatic of changes taking place in the way in which development is perceived and therefore in development planning itself. The most recent Unesco publication in this area, "The use of socio-economic indicators in development planning", edited by Nancy Baster, The Unesco Press, Paris, 1976, summarizes this, pp. 9-10:

"A shift from human resources indicators to social indicators. This means that cultural and educational indicators are no longer seen only as resource indicators within a basically economic model, but as social indicators within a broader socio-economic model. This introduces greater flexibility but at the same time leaves open the nature of the interdependence between economic and social variables.

A shift from social to socio-economic indicators. There seem to be at least two elements in this shift. Social indicators are closely linked with economic indicators and both are needed in explanatory models. Integrated systems of socio-economic indicators are being sought both from the social side and from the economic side. Until recently, this has tended to be a rather one-sided affair with the social variables playing a subordinate rôle in economic growth models. New changes in economic models, and particularly pressure for more disaggregation in economic accounting, is

creating a demand for a wider range of economic as well as social indicators. This makes possible a much more open-ended analysis of patterns of socio-economic development, but again leaves open the theoretical framework.

A shift from the use of indicators to measure international differences, to the use of indicators to measure the socio-economic situation and trends within countries, with particular emphasis on social, economic and spatial differences. This is part of a move away from universalistic prescriptions to a kind of development planning which gives greater weight to the actual historical pattern of development in particular countries. It is also part of the distrust of national aggregates, and the pressure to look beyond them to the pattern of distribution and dependency, and to problems of inequality, poverty and unemployment.

A shift to the operationalization of indicators. In recent years, a great deal has been written about social indicators and a number of 'systems of indicators' have been proposed. How far are these applicable and useful in planning? How, and in what way, do planners use indicators? What is the relation between the development of indicators in order to analyse the interrelation between social and economic variables in the process of development, and the use of indicators in planning and programming?"

A number of studies on indicators of economic and social change have been undertaken since 1973 and two informal meetings in this area were held at Unesco Headquarters, October 1973 and March 1974. A set of studies on the use of socio-economic indicators in development planning in Asia was carried out and discussed in a regional meeting of experts held in Bangkok, November 1974. A number of the studies carried out under these projects up to 1974 are contained in the book edited by Baster.

In April 1976, the first meeting of experts specifically dealing with indicators of social and economic change was held, which had before it a series of papers on theories of development and change, indicators of development and change, and their use in planning in selected countries. One of them, "Human needs, human rights and the theories of development", by Johan Galtung and Anders Wirak, constitutes Part I of this issue.

The regional meeting of experts in Asia was followed in 1975 by two in-depth case studies, one in the case of the Philippines, and the other in the case of Thailand, on the application of territorial indicators as an input in the development planning process, which were discussed at national workshops held in Bangkok (Thailand), and Manila and Iloilo (Philippines) in January 1976, which brought together a large number of researchers, planners and policy-makers in those countries. The two case studies form Part II of this issue.

The intention, therefore, is to illustrate the conceptual and practical aspects of this programme. These studies are but two of many available in the Division. A bibliography is appended.

Contents

Part I

Human Needs, Human Rights and the Theories of Development*

Johan Galtung and Anders Wirak
Chair in Conflict and Peace Research, University of Oslo;
Inter-University Centre of Post-graduate Studies, Dubrovnik
Institut d'Etudes du Développement, Geneva

Contents

The views expressed in this paper are those of the authors and not necessarily of the Organization.

---

* Unesco Workshop on the Applicability of Social Indicators to National Planning in Thailand, Bangkok, 23 January 1976. Document No. SHC-75/WS/55, 20 January 1976.

I. INTRODUCTION
THE NEED FOR A NEEDS-ORIENTED
DEVELOPMENT THEORY

The purpose of this paper is to discuss theories of development, with a particular view to the problem of how development is to be measured. In other words, the goal is to somehow formulate development theory in such a way that it can be tied to a reasonable battery of social indicators.

In order to do this we shall start with one very simple premise: the point of departure will not be existing social, economic and/or political indicators - we shall not permit them to prefigure our thinking about development. Indicators are tools, but they are not innocuous tools. They have built into them assumptions about how development is to be conceived, and the moment one lands on one indicator a number of questions has already been implicitly answered - very often without the knowledge of those who engage in the exercise. Thus, it can probably be argued convincingly today that the Gross National Product is above all a measure of a particular way of organizing the production system, based on a high level of processing of raw materials and a high level of circulation of factors and products, including marketing. Thus, the indicator reflects industrialization and market and mobility-oriented economic systems, both of them combined in present-day capitalism, whether the decision-making is primarily made by privately-employed or State-employed managers. But this is one of several ways of organizing the system of production and consumption. Another way would be based on a low level of processing in the direction of the satisfaction of basic material needs (food, clothes, shelter, health, education), and relatively limited economic cycles (production primarily for local consumption, consumption primarily of local products). Obviously, a system of that kind would rank low in terms of degree of processing and degree of circulation, marketing, and might tend to be identified with the traditional economies development is supposed to break away from. Why bring them in?

Because there are some reasons that have come to the foreground in recent political theory and practice. First, there is the experience that even societies undergoing rapid economic growth as measured by GNP per capita often seem incapable of satisfying even the basic material needs at the minimum level for vast proportions of their population.[1] Second, there is the circumstance where economic systems that aim more directly at satisfaction of basic needs at the same time may also satisfy a number of basic non-material needs that often seem to be left unsatisfied, even counteracted by certain factors of societies in rapid economic growth.[2] Needless to say, all this has engendered a fresh debate on development, and a search for some type of new point of departure. A new set of indicators should grow out of that search, and in that set some of the old indicators might be included, but then on the basis of a well thought-through rationale.

Second, it is another basic assumption of the present paper that most of our present thinking on development is marred by one particular intellectual fallacy that is committed over and over again. The fallacy is the following: instead of conceiving of development as development of man and woman everywhere, development has been conceived of as the development of things, systems and structures. The view that will be taken here is that these are all means; that the very purpose of the process of development must be somehow concerned with human beings, not to improve the quantity and quality of things. Systems and structures have to be changed, but that is not the goal of development, only the instruments not the end, and at most the means.[3] More importantly, the acid test of whether development has taken place or not is not located in the development of things, systems and structures, but in what it does to human beings.

This formulation may sound general and even tautological, but when looked at more carefully is full of implications that will be followed up throughout this paper. First, it follows from this that there are no "sacred cows"; economic growth, democracy, socialism, may all represent important approaches in the field of development in the sense of contributing to the accumulation of things, changes of systems and changes of structures, but they all have to prove themselves. The ultimate test lies in what they, concretely, do to human beings. One might say that consequently, all that really matters are human beings themselves.[4]

Thus, this immediately leads to the question of what could possibly be meant by "development of human beings" - or, in another terminology, "personal growth".[5] This will be discussed in a later section. Here it suffices only to say that the question opens an area of thinking and research hitherto unrelated to development studies. In a sense this follows from the basic assumption that the theory and practice of development have concentrated on something else. To get some kind of image of possible foci of development theory and practice, for the purpose of this paper, let us make use of the simplest possible distinction, between MAN-SOCIETY-NATURE, and for our purpose distinguish between the three aspects of society mentioned above: the things (in a broad sense, anything that is produced), the system (here we are thinking particularly of the system of distribution) and the structure (here we mean the whole set of interacting relations, bilateral and multilateral, to be found in any society.

Thus, we arrive at the following figure:

Figure I

The foci of development theory

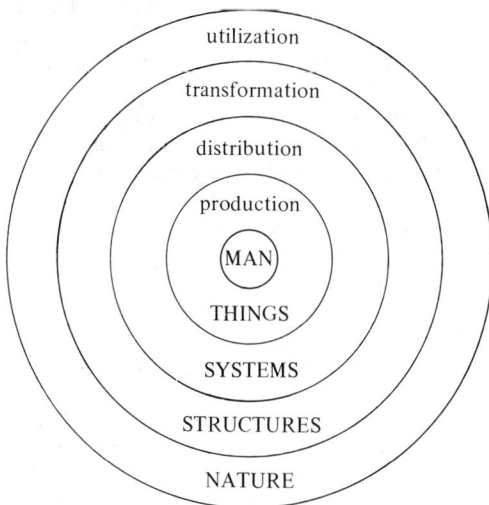

Some comments on this figure.

One might now summarize what has been said so far in the thesis that the most important development theories in the post-Second World War period (to be analysed in more detail in the next section) have focused on the three intermediate circles, to the exclusion of a real study of Man (what his needs are and whether what has taken place under the name of development really has served Man well,[*] and to the exclusion of Nature. The focus has been on how one can utilize Nature for production and distribute what has been produced, with or without structural transformations. In recent years, however, the idea of outer limits has become more and more prominent: there is a limit to how far we can pollute and deplete. With this a new focus on the development debate has come into play, and much more vigorously than any new concern for the development of human beings as such, or "personal growth".

Having mentioned this about the "outer limits" of Nature it is tempting to introduce the other concept used by the United Nations Environmental Programme: the "inner limits". These limits are supposedly located inside Man and they can be conceived of in terms of satisfaction of human needs. Just as Nature sets outer limits beyond which Man cannot go in his utilization of Nature without being destructive not only to Nature but also to himself, the inner limits are those below which human needs cannot be satisfied without fundamental damage to Man. (6) This states the developmental process rather neatly: it is a question of conceiving of those processes that pushes human need satisfaction above the inner limit, without at the same time pushing the utilization of Nature beyond the outer limits.

Having said this, it becomes once more important to underline how lopsided the theory of development has tended to become. If one tries to locate development as a process striving to serve Man in the context of our shared environment, then Man and Nature have to be brought into the centre of theory formation and developmental practice; not be made peripheral, as constraints or "Randbedingungen". And that in one sentence is perhaps the purpose of this paper.

II. SOME CHARACTERISTICS OF CURRENT THEORIES OF DEVELOPMENT

By and large, the position to be taken in this paper is that there are two classes of theories of development of any importance in the world today, broadly termed the "liberal" and "Marxist" approaches. There are both differences between them and similarities. We shall start with some of the similarities.

Both are very concerned with production; with the whole economic cycle from extraction from Nature, through the various levels of processing, distribution to consumption - including consideration of the whole superstructure that comes along with these elementary and fundamental aspects of production, such as financing, research and development, administrative structures and with the whole nation-State structure and international superstructure built on top of that again.

Thus, both of them are eminently economistic in their focus, and in the conflict between liberal and Marxist economists or "productionists" there might well be a hidden underlying alliance in the interests of preserving the predominance of economism-productionism as an approach. It should be pointed out that this so far has had a tendency to leave out the two concerns that were voiced in the introduction: the concern with the outer limits of Nature (and here the Marxists have "sinned" as much as liberals), and the concern with a deeper theory of Man.

The opposite of a deep theory of Man is a superficial theory of Man, and one way in which apparent antagonists can meet, on relatively shallow ground, is to focus the theory and practice on basic material needs only. Nothing of what follows is in any sense a denial of the fundamental importance of those needs; to the contrary, as will be clear from the next section they should certainly be seen as focal, basic in any theory of Man - but not as the only focus. The basic material needs most often emphasized are the ones mentioned

_____

[*] We apologize for this "male chauvinist" language - not being responsible for the inability of the English language, and other languages such as French, to express in one word man + woman. "Human beings", "persons", "people", all three have somewhat different connotations.

in the preceding section: food, clothes, shelter, health and education. In saying this we are not denying that Marxist thinkers and liberal thinkers in general have much richer concepts of man than what can be subsumed under those five simple headings, but when that thinking is narrowed down the way economists are wont to do, and more particularly channelled into development theory and practice, the richer connotations tend to get lost.

Of course, liberal and marxist theories of development will tend to differ, very basically, as to the means of satisfying the basic needs. This theme will be developed shortly; let us only pause here to reflect on why the leading schools of development should be similar as to focus of attention.

One approach to answering this question would be to take as a point of departure the rather obvious circumstance that both schools of thought were developed, essentially, by people who were Western, intellectuals and men, and these schools grew out of a period very much concerned with industrialization, capitalistic economic growth, nation-State building and empire-building. The schools of thought certainly have different views on these matters, but since they represent different answers to some of the same problems and are also inspired by the same empirical context and the same fundamental paradigms (ideas as to what constitutes the ontological unit) it is not strange that they should end up with great similarities.

Thus, one could imagine that if development theory and practice were not so much in the hands of Western intellectual men the following alternative characteristics might have been more prevalent:

(1) More concern with the immediate human situation, with how a person feels, lives and loves, acts and behaves, with such things as happiness, feelings of self-realization, having a sense of purpose, and so on. It should be pointed out that thinking in terms of things, systems and structures is at a higher level of abstraction, more removed from people in general, and hence something that serves better the interests of intellectuals. It can be done in terms of abstractions, narrowing a complex reality into a set of units and variables and making experts out of people who are able to manipulate symbols of that kind. Hence liberal and Marxist intellectuals meet in the concept of planning.

(2) Both schools have tended to take industrialization for granted. In other words they have not differed essentially when it comes to the idea that much processing is needed to produce things. (7) (The difference lies in the views on system distribution and structural transformation.) Since they have tended to take industrialization for granted they have not been concerned with asking such questions as limits to industrialization (which is not the same as the more general question of limits to economic growth). The negation of that type

of question has been facilitated by the narrow concept of man, narrowed down to a few lines in the full spectrum of human needs (such as the five basic material ones alluded to above), and it has seemed beyond doubt that some type of industrialization is needed in order to produce food, clothes, shelter, health hardware (medical equipment, etc.) and education hardware (schooling equipment, such as textbooks, buildings, etc.). Again one might sense something Western in the selection of a few physical variables to the exclusion of many others, and in the reduction of extremely complicated nature down to a more manageable set that can be handled somewhat in the same way as, for instance, a car is managed.(8)

(3) Another fundamental similarity lies in the concern with the nation-State as the unit of development. Characteristically indicators of development are calculated on a national basis, the nation-State being the fundamental focus, that which "develops". For the liberal the State is the scene of economic growth and distribution, for the Marxist it is the stage of revolutions. Hence both will tend to conceive of the country as the unit of development, the country thus being a collection of "things, systems and structures". Correspondingly, the problem of development can be fitted into a paradigm that sees the world as essentially composed of States, in co-operation and conflict, making much of development thinking isomorphic to strategic analysis. (9)

Again, we do not want to deny the relevance of much of what has been alluded to in the three points above, but only want to point out that here there are common biases that may unite contending factions and make conflicts between them look like the famous tip of the iceberg. But, there are, of course, also very real differences in how they look at the development of things, systems and structures - in other words production, distribution and transformation.

As a first approximation let us present the difference as follows:

The Liberal View: increase production, then distribute; make the social transformations that are necessary in order to increase production and derive information as to which transformations would be conducive from the societies which have been able to achieve the highest level of production.

The Marxist View: fundamental social transformation of the structure first, increased production and better distribution as parallel processes that would be made possible by the transformation later.

One might say that the liberal view is in a sense an optimistic one, stating that it is possible to progress without any basic transformation of society except that the less developed societies should learn from the more developed ones. The Marxist view questions this, and would, initially, distinguish between countries in the centre and in the periphery of the world capitalist system, the metropolises versus the satellites, the dominant versus the dominated, the autonomous versus the dependent countries. Some remarks are called for on this since no discussion on development is possible without bringing this aspect of the contemporary world into focus - under the heading of structure, global and domestic.

The particular social economic and geopolitical system known as capitalism has to be brought in in this connexion. Needless to say, liberal economists have tended to regard the basic assumptions in the capitalist way of organizing production and consumption as somehow being natural; (10) Marxists have rejected this and have seen capitalism as one among several possibilities, contrasting it with feudal and socialist formations, in the Marxist theory of successive stages. (11) Liberals have seen development as something that takes place basically within the capitalist system; Marxists have conceived of it as something predicated on the assumption of moving up from the capitalist system into higher stages. The difference is certainly fundamental, but there is also a similarity which is profoundly Western: the idea of progress in the continuous, accumulative system - immanent form of liberal theory as well as in the discontinuous, non-accumulative and system-transcending Marxist version.

To get at this similarity, an image of what capitalism is about is needed, and the one presented here is an effort to combine some liberal and some Marxist perspectives on capitalism. Capitalism is thus seen as a way of organizing production (and consumption) with four major characteristics:

(1) Capitalism is capital centred, meaning that the criteria of whether an economic process is a success or a failure can be measured in capital terms (e.g. increased assets, increased sales, etc.). This should be understood relative to alternative criteria of success, e.g. whether the process enriches and ennobles those who engage in the work (human factor-centred), or enriches the nature from which the raw materials are extracted (nature factor-centred), or is simply product-centred (focusing on, for instance, to what extent five-year plans or quotas are fulfilled or unfulfilled). Profit-motivation is compatible with this.

(2) Capitalism is based on division of labour between those who own means of production and those who do not (the labour buyers and the labour sellers), and between those who define and solve problems (decision-makers, researchers etc.) and those who implement the solutions (the supervisors, the workers). In earlier stages of capitalism, this dividing line was linked to the caste-like dividing lines of feudal society, in later (present) stages, it is correlated with education levels which in turn may be correlated with residual caste aspects of modern societies.

(3) Capitalism is mobility-oriented, meaning that the system is based on a very high level of mobility of the factors of production to the points where they are processed, and from which they are marketed. In this process raw factors are taken from regions of a country, or from countries in the world, in the form of raw materials, raw capital (e.g. savings) and raw labour (unskilled workers and untrained "talents") and brought to points where these raw materials are processed into products, the raw capital into finance capital and raw labour is made use of and/or is processed into more skilled types of labour. This mobility creates very steep centre-periphery gradients within and between countries, between the centres of processing where the factories, the financial and educational institutions are located and the peripheries from which these raw factors are taken.

(4) Capitalism is expansionist, meaning that the mobility described in the preceding point knows no political borders, but at every time will always try to transgress whatever borders there are, thus constituting a pattern characterized by its global reach. (12) The epitome of all this would be the current multinational corporations for economic activities, tying together as vast portions parts of the world and as many segments of the economic cycle as possible under one administrative, integrated leadership.

These four characteristics are very often seen as being closely related to the development process itself which is then described in terms of capital formation, the growth of a social formation corresponding to this particular way of organizing production with an upper class to command the various types of processing described above and middle and working classes to carry out the decisions, with a high level of processing of raw materials, raw capital and "raw" human resources by means of factories, financial institutions and schools at primary, secondary and tertiary levels, in absolute and relative terms. It should also be pointed out that social-economic "growth" has had as its correlate, partly as a cause, partly as a consequence, a corresponding process of political growth in the formation of the modern State. A modern State is in administrative/political terms what capitalism is in social economic terms: a centre-periphery gradient built into geography, with its centre in the capital, its periphery elsewhere and particularly towards the borders (generally speaking), and with a high level of division of labour between what happens in the centre and in the periphery. Thus far we have mainly mentioned what the periphery sends to the centre (raw production factors), but the periphery also receives something from the centre: finished products for consumption, capital for investment, and human labour in processed form, particularly in that of functionaries. When this system creates an overspill across political borders "peripherizing" other countries it is known as colonialism; the corresponding overspill in economic terms is known as (capitalist) imperialism.

This is not the place to discuss the development of Western political colonialism and economic imperialism for the last five hundred years, roughly speaking, since Columbus and Vasco da Gama sailed westward and eastward, respectively from the Iberian peninsula. (13) What is important here are only the many consequences for development theory, some of which we shall mention here.

First, Western expansion has been so thorough that up to recently it has been taken for a normal process, and Westernization has been confused with modernization and development. Of course, Western expansionism did not necessarily have to lead to Westernization. It is also possible to extract raw materials, raw capital (by the very practical method of taxation) and raw labour (by means of deportation) without creating homologous countries in the periphery. Occupation with well-distributed garrisons might be sufficient for robbery and taxation. But it seems somehow to be embedded in Western expansion, perhaps as opposed to expansionism engaged in by other cultures, that this is not enough, and that Western social-economic and geopolitical formations constitute model patterns making it not only the right, but also the duty of the West to engage in Westernization. (14) For a long time, the assumption seems to have been that the West not only possessed model societies, but also model creeds, model scientific methods, model production techniques, etc. Thus, a concept of development came into being whereby development was, grosso modo, the degree of proximity to the standard Western model and underdevelopment was measured in terms of distance from it - unidimensionally in the more simplistic cases (the GNP per capita or simplistic economic growth school being one example), multidimensionally in the more sophisticated cases.

Second, this in turn led to the unilinear school of development thinking, with the Western society as model society and in the post-Second World War period with the United States playing the rôle of the "best" model. Development indicators tended to become those by which the United States would come out as number one; taking its model quality as axiomatic. Ideas of "catching up", "narrowing the gap", or "process of development", all of them relating to the unilinear idea, caught hold. What was not questioned, or has only been questioned recently were such items as:
- would not imitation of the model country also lead to the growth of the less desirable aspects?
- in order to imitate the model country would one not have to engage in some of the same methods, and would that not presuppose that one would have to make use of other countries as a periphery?
- was it obvious that the Western model could hold trans-culturally, or might there be cultural differences of different kinds that would make the Western model untransferable?
In other words, it was not only the question of whether it was possible to transfer the Western

model, but also that of whether it was desirable which tended to be overlooked.

Third, curiously enough this also led to an important weakness in Marxist thinking. The first country to break away from the capitalist mode of production was the Soviet Union after the October Revolution in 1917. The breakaway took place in the periphery of Western capitalism, not at its centre as predicted and expected. This seems to have had two major implications: the Marxist theory of development has above all become one for periphery societies of the capitalist system, which roughly correspond with what liberal thinking would identify as "developing societies" (in terms of such absolute properties as GNP per capita, level of industrialization and education, etc., rather than such relational properties as how the country enters in the international division of labour, level of dependence, etc. that Marxist thinking would focus on). And as importantly: out of the Russian periphery of Western capitalism came the mighty power of the Soviet Union, also a Western country - it seems - to some extent dominated by the idea that it is a model country, establishing its own unilinear track of development for its own periphery; formerly that of capitalism.

Thus it was that the period after the Second World War came to be characterized by two unilinear tracks of development: one based on a Rostowian theory of stages of economic growth with the United States as model country, and the other based on a Marxist theory of stages of social formation, with the Soviet Union as the model country. The two tracks would seem to be at cross purposes with each other, and so they are or have been in some countries, but they also are remarkably parallel. In both cases it can be and has been proven rather convincingly that it also serves the interests of the model country if its periphery accepts the theory of development. In other words, a periphery that tries to become as much like the model country as possible will also be a better one in the sense that the centre not only can more readily establish a division of labour with it (making use of the bridge-heads established in the centre of the periphery countries), but also in the sense that the centre countries, through striving of the periphery countries to do their best and progress along the tracks of development find a validation or reconfirmation of their own systems and structures.

We have gone into this at some length because, like any other social theory, theories of development have to be understood in their social context. What we have said can be summarized as follows. Out of the richness of liberal and Marxist thinking there has been a tendency to crystallize theories of development that have created a development thinking and practice leading to results not only compatible but sometimes identical with the basic structure of power relations in the world today.

This is also important in understanding the material out of which alternative theories might grow and these also have to be seen in a general social context. This might serve partly to explain the importance of the position held by China today as the model of development for many in many countries. It is the only country of significance that has rejected both unilinear tracks. In so doing one might even say that they have been rejecting not only the idea of the United States and the Soviet Union as models, but the more general idea of the Western world as constituting a legitimate model area for general imitation. It would then be in line with what has been said above if the net outcome is the establishment of a third unilinear track with China as the model country. Although one hypothesis might be that China, like her Asian neighbour Japan, is not bent on being imitated, but rather, that there may be other countries which have been under Western influence that are bent on being imitators. (15)

III. THE LIBERAL AND MARXIST DEVELOPMENT MODELS: SOME DETAILS

Against the backdrop of what has been said in the two preceding sections, let us now start to look in somewhat more detail at the basic differences in liberal and Marxist development theories, given the similarities in the emphasis on industrial production, and the weakly-developed concern with Man and Nature.

One way of approaching the differences is contained in Figure II:

Figure II

Liberal vs. Marxist development theory: two axes of development

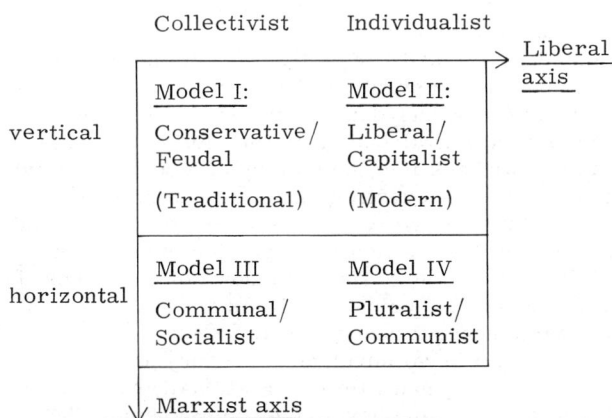

	Collectivist	Individualist	
			→ Liberal axis
vertical	Model I: Conservative/ Feudal (Traditional)	Model II: Liberal/ Capitalist (Modern)	
horizontal	Model III Communal/ Socialist	Model IV Pluralist/ Communist	
	↓ Marxist axis		

Some comments on the definition of the two axes follow.

The horizontal or liberal axis is a dimension of increasing individualism, both in the sense of inter-individual diversity and in the sense of individual mobility. Translated into European history, what it purports to harbour would be something like the transition from mediaeval society into the early modern period, with the Renaissance as the transition period. Imagine we agree with the idea that this transition period signified some kind of release of the individual from the captive cocoon of mediaeval/feudal collectivism, both in the sense of cultivating diversity (with creative originality also under that heading), and in that of a much higher level of individual mobility. There is, however, another factor alluded to above that should also be emphasized: among other things this mobility expressed itself in the travels from Europe to the West and to the East, and hence in a new geographical scope to Western latent expansionism. Individualism and expansionism coincided in time.

The other or the "Marxist" axis is concerned with a completely different emphasis in the whole conceptualization of development. The dimension has to do with increasing horizontalization in society, both in the sense of decreasing differences in power and privilege, and in the deeper sense of a more horizontal division of labour; one in which the benefits, including the psychological benefits in terms of challenge, access to stimulating work, are more evenly distributed between the partners in a production process. Production should then be taken in a broad sense: it might, for instance, include the production of art, and the question of how readers/viewers can participate in the production process as co-producers with the artist. Again there is a possible image of two types of societies, one being more vertical, the other more horizontal with a transition period initiated by the Russian October Revolution of 1917. One might here imagine that Marxists, except the more dogmatic ones, would talk about transition periods rather than the short span of a couple of years or one generation in, for instance, Soviet history: a period lasting at least as long as the Renaissance, including a variety of phenomena such as the individualization of the religious experience, Protestantism, which can be said to be a part of the Renaissance process as conceived of above. However, many Marxists have tended to conceive of horizontalization in an extremely narrow manner, in terms of something happening to the ownership of the means of production - in line with the general inclination towards economism mentioned above - compatible with thinking in terms of transition points (revolutions), rather than periods (evolution).

It is now our thesis that development theory in general gains from the inclusion of both axes, and that the liberal and Marxist schools have had a tendency to over-emphasize one of them at the expense of the other. We shall now elaborate on this point.

We assume that the basic tenet underlying the entire liberal exercise is faith in the individual. The entire view is profoundly actor-oriented, seeing development largely as the creation of particularly capable and active individuals. The individuals are the entrepreneurs, in any field: economic production, politics, military matters, culture. By and large they are the engines pulling the train. For them to function two conditions have to obtain: they have to be given sufficient freedom of movement to carry out their initiative, and they have to be sufficiently motivated by the expectation of a reasonable reward. This reward system has to be differential, or else it would not distinguish between individuals where talent is concerned. And this differential reward is the social verticality, which then stands out as one of the conditions for development.

In other words, the liberal theory of development would essentially be concerned with the transition from Model I to Model II in Figure II, or from conservative/feudal to liberal/capitalist societies, but in development theory of the 1950s and 1960s was largely referred to as a transition from "traditional" to "modern". Verticality being taken for granted, in general terms, does not mean that its mechanisms do not change. From rewards being roughly proportionate to one's station at birth ("like father like son") we move on to rewards commensurate with innate talent, leading in turn to the need for measurements of such talent (this is where Binet and Simon enter with the idea of I.Q. tests), and the institutionalization of the correlation between ability and reward. This institutionalization is a system of cumulative schooling through stages of primary, secondary and tertiary education whereby a new élite is formed and legitimized. The system is then purified under the slogan of "equal opportunity", meaning that there should be no carry-over from Model I society as to who shall be given the highest level of educational legitimacy; that being decided on the basis of ability alone. This idea more or less sums up the nature of the debate between conservatives, liberals and social democrats within Model II society, a debate not concerned with verticality as such.

So much for the first aspects of individualization: the detection (through a grid of schools) and cultivation (through processing of individuals by means of successive levels of schooling) of the capable and active. The second aspect was the idea of individual mobility, which also has played a fundamental rôle in the liberal theory of development.

When, for instance, the leading U.S. sociologist in the post-Second World War period, Talcott Parsons, tries to describe development in terms of a transition from a pattern characterized as particularistic/diffuse/ascribed to universalistic/specific/achieved, he manages, admirably, to say in few words something very basic about at least the non-élite aspect of the transition from traditional to modern. (16) This transition period has also been characterized by the growth of capitalism as a mode of production and by a high level of mobility of factors, including human labour. What Parsons expresses is already contained in the very idea so clearly expressed in economic thinking, of treating human beings as a "factor" to be handled analytically in the same way as raw material and capital. Mobility is facilitated by substitutability, with modern physics/chemistry/geology/biology establishing a level of abstraction whereby minerals, animals, plants etc., can be seen as belonging to the same class and, hence, be interchangeable because they share basic characteristics.

What happened in the period of transition characterized by the pattern variables employed by Parsons (actually taken from the work of Sorokin's monumental Social and Cultural Dynamics) was that institutions emerged so as to make human beings also more interchangeable. Universalism stands for the idea of inter-subjectively communicable and accepted criteria of evaluation of another person - and in modern society this is the input of schooling to society in the form of diploma; constituting a universally accepted basis for evaluation as opposed to, for instance, kinship relations (not universal, the person is my nephew, not everybody's nephew), or personal sympathy (not a universal relation either).

Correspondingly, under the heading of specificity there is the important idea that the evaluation shall be relatively unidimensional, otherwise it becomes so complicated that universalism is threatened. In modern societies, schooling also provides for this through a system of evaluation of a particular type of intellectual ability, with a high component of memory and a special type of behaviour fostered in schools in general, and on examination days in particular. Then finally, under the heading of "achieved" rather than "ascribed" comes the whole idea that the individual has to prove himself; what he inherited at birth, such as the position of the father shall wash out and the individual shall emerge. Schooling also plays a rôle here in organizing achievement exams whereby individuals are sorted and ranked vertically, but this is also done in most other institutions in liberal society. Hence, what Parsons has done is to put in sociological terms what had already happened for a long time (dating far back into the history of Model I society too): the easy substitutability of one individual for another so that one knows where to look for the substitute if a position is empty due to social or biological demise. The breakdown of the extended and even of the nuclear family is another aspect of the same general phenomenon.

Thus, liberal development theory has above all been concerned with the irrationalities of traditional society, and how to overcome them. The present critique of what we have referred to as Model II society might perhaps have changed our focus somewhat and perhaps made us more able to see that what is irrational from the point of view of a particular economic formation may be highly rational if the point of departure is a richer view of human beings. Thus, when a chemist/mineralogist tells us that a mineral has the formula $CuSO_4$ (copper sulphate), everything which when exposed to certain tests shows the same characteristics becomes equivalent; a chemical formula establishes an equivalence/class of mutually interchangeable objects. But it so happens that one unit in that class may have tremendous beauty, another may (be said to) possess magic qualities, while still another may relate to a person due to some kind of emotional experience, etc. Traditional societies may institutionalize the particularities which disappear in the light of Western science, and reappear as obstacles to development when seen from a point of view that takes modern society for granted, even for "normal". Similarly with money: it is well known that economists have been able to build a mathematical theory where cost, price, capital value play fundamental rôles as variables. The mathematics used is powerful, for instance, because it can be based on such properties of certain mathematical systems as the commutative and associative laws:

Commutative law: $a + b = b + a$

Associative law: $a + (b+c) = (a+b) + c$

The Western modern mind has been trained in such a way that not only are these two principles recognized and taken for granted by most who have been through the schooling process, but so are their applicability to money. But the implication of the commutative law, for instance, when applied to money is that the order of payment given and payment received does not matter; it all adds up anyhow. If two people, A and B, owe me money, it does not matter to me whether A or B pays first. But that is a social cultural dictum, and certainly not generally valid, for instance in most - one might assume - traditional societies. In accepting such mathematical principles as valid for money, the social decorum, embroidery so to speak of many traditional societies, is cut off and the rest is very plain cloth -- albeit very much of it produced by modern methods.

The most dramatic implication of this comes with the human factors: inter-human substitutability. It is trivial today that this has been brought about in the production process through the transition from artisanal to industrial mass production. The point is not that this was also a transition from labour-intensive to capital- and research-intensive methods of production; away from such methods of production as building a dam with picks and shovels, baskets and naked body power alone. The point is that this was also a transition towards a much higher level of substitutability, meaning that the individual worker had much less of a chance to imprint the finished product with his own personality. (18) The intellectual has this possibility, the present authors are for instance, relatively convinced of their own unsubstitutability in the sense that nobody else could produce exactly this present

paper. But for most people in modern society, this direct relation between worker and product is eliminated. The worker is detachable from his or her product, substitutability has been institutionalized. Needless to say, this is related to the Marxist concept of alienation - without any pretence that it captures the richness of that concept. More particularly, it should also be pointed out that here alienation, or substitutability is seen as an objective category, built into the social structure. How it may or may not be reflected in the individual psyche as unhappiness, boredom, stress, possibly neuroses and psychosis, including schizophrenia, or delight at being so detached from work that one can concentrate fully on leisure-time activity is another matter.

Thus, one may say in general that liberal theory has built into it the verticality of liberal society: it deals very differently with élites and masses. Elites, duly legitimized through education and/or other achievements, are granted full individuation in the sense that they can make themselves unsubstitutable. They then become model individuals to be emulated, admired and envied, loved and hated. The masses are not given that chance, "theirs is but to do and die". For them, maximum substitutability is prepared institutionally. Universalism, specificity and achievement are for them. At the top of society, particularism (who knows whom), diffuseness (multiple relationships, not only in business, also in politics, in leisure-time activities, on committees and boards) and ascription are the order of the day (someone who has achieved in one field and surfaced on the top of Model II society as an individual is made use of in other fields and higher up, often totally beyond his or her competence).

All this is then correlated with division of labour as described above in such a way that it is given to the élite not only to try to solve problems but also to define what is an acceptable problem, and to the masses, in a substitutable manner, to implement the blue-prints emanating from the problem-solving activities of the élite. Thus, the masses become the supernumeraries with the élite centre stage (one stage being the mass media), or the spectators of élite activity. The élite legitimizes its position through the "trickling down" hypothesis, meaning that in the shorter or longer run the benefits from élite activity will accrue to the masses, for instance, in the form of economic growth well distributed or general, mass education.

There seems to be little or no doubt that a Model II variety society as described here can engage in economic growth, often for a sustained period of time. Many of the development studies in the 1950s and 1960s were devoted to showing the correlations between the various aspects of Model II society and economic growth, the correlations all tending to be positive, some of them highly so. (19) This holds both for synchronic and diachronic relations, both when tested for many societies at one point in time, and for one society over a time period.(20) But, as any methodologist would warn, any correlation can be spurious. Economic growth may be due to something else occurring at the same time as the type of change described above also takes place and that something else may also be underlying that other type of change. The question then becomes what could that something else possibly be?

One possibility would be the increasing domain and scope of the centre-periphery gradient that come with capitalistic modes of production. Thus, the argument would be, and this is of course central to modern theories of imperialism, that economic growth in a centre, Model II type country has been made possible only because it has extracted factors from a vast periphery, and succeeded in putting most of the periphery outside itself, in the patterns known as colonialism and neo-colonialism. At the same time the transition from Model I to Model II has taken place, but this is hardly crucial; what is crucial may be the availability of a malleable, penetrable periphery. Thus, Japan cannot be characterized as a Model II society even today. It still has strong Model I features given the way in which level and type of schooling are used ascriptively, and the way promotion is correlated with seniority (another ascribed characteristic, although universalistic), rather than with achievements. But Japan certainly established a periphery starting with the Sino-Japanese war from 1894 to 1895, and continuing until after the 1945 defeat with more neo-colonial methods.(21) Besides, it may also be argued that the real function, intended or not, of liberalization of society, of equal opportunity, and of mobility in and out of classes destroying the caste system of Model I society has been to create societies in the centre where the classes have a shared interest in maintaining control of the periphery. If there is something to this type of reasoning, periphery resistance should be highly consequential. More concretely, decreased availability of a periphery for the extraction of factors, and as markets for products should reduce economic growth considerably, while simultaneously mobilizing all segments of society in its struggle to preserve the centre periphery gradient, because of the common interest in "sharing the spoils". Recent events, after the "energy crisis", point in this direction.

In saying this we have in a sense already proceeded to the third circle of development thinking: the distribution within the systems. There is no doubt that liberal theory, particularly in its social democratic variety as it is expressed in Northwestern Europe, is strong in its emphasis both on social justice and on equality. One example of social justice, the idea of equal opportunity, has already been considered earlier. This should not be confused with equality; it only means an equal chance to participate in the competition for

positions in a vertical society. Liberal theory would, however, also be strong on other types of social justice which can all be summarized under the general formula that there should be little or no correlation between ascribed and achieved status. Thus, whether you are male or female should not influence your level of education or position in the power system at the micro or macro level; whether you are born in the city or in the countryside shall also be irrelevant, not to mention the axiomatic irrelevance of race and ethnic factors. According to liberal theory a society could hardly be considered developed on the basis of economic growth alone if gross social injustices of the type indicated above were still prevalent.

However, the formula for social justice just given is an open-ended one. One has long been concerned with racism, but only very recently with sexism, and still practically not at all with "ageism" - defined here as the institutionalization of the rule of the middle-aged over the very young and the very old, relegating them either to ghettos (kindergartens and schools; old-age homes and retirement), fragmenting them away from one another. And if, as some liberal theory may have it, it is true that there is such a thing as relatively steep differentials in innate ability, for instance, in that by measurement which is attempted through I.Q. tests, then this would also be seen as an ascribed characteristic and hence lead to a movement for the abolition of discrimination against the less talented. These points are made here precisely because of the open-ended nature of the social justice idea, as well as to indicate that the idea is very rich in its implications and will keep Model II society busy both with theory formation and with political practice for some time to come.

The issue of equality is a different one. It is a question of abolishing differentials in power and privilege, very often approached in terms of income distributions. One could also approach it in terms of other dimensions of "having", such as the distribution of all kinds of consumer goods, food, clothes, shelter, access to health and education facilities, to transportation and communication, etc.

The trouble with this approach, of course, is that it locates the equality problematic in the consumption part of the economic cycle rather than the production part, and within the economic frame of thinking. The latter, however, has certainly been transcended in liberal thinking. There is the old tradition of seeing development not in terms of institution-building for better control of the periphery (that is also included), but in terms of power sharing, starting with the citizen concept and expansions of the right to vote at the macro level, proceeding to some type of decision sharing inside enterprises, today known as industrial democracy. We mention all this because it would be totally wrong to assume that liberal development theory is only concerned with economic aspects although the point of gravity has been in the economic growth field. Stability based on balance of economic power (anti-monopoly), political power (party formation), military power and cultural power (pluralism) is also part of the theory.

However, although trade union formation and in general organizations of the underprivileged are important in overcoming fragmentation, based as they are on solidarity. And although voting rights, etc. are important in overcoming marginalization, creating a more generally participatory society, the basic verticality relating to the production aspect of the economic cycle still remains untouched.

And this is, of course, the point where the Marxist onslaught starts. Like liberalism Marxism is a set of theories, and similarly one can find almost any statement if one looks at a sufficient number of texts sufficiently exegetically. Our presentation here will also be critical, not in any sense doing full justice to the richness and depth of Marxist thought, which in its fundamentals is very different from liberal thinking (the difference partly being absorbed in the word pair dialectics vs. positivism), focusing more on its expressions in development practice today.

From the point of view of development theory the strength of Marxist thinking is in its analysis of the structure of economic production. The Marxist critique operates on at least three structural levels: inside the production unit (the firm); domestically, within the country and at the global level. In general terms it may be said that the Marxist criticism takes up in considerable detail the last three of the four characteristics of capitalism mentioned in the preceding section: the division of labour between human beings inside the production unit (the firm), between those who are problem-solvers and decision-makers and those who implement solutions and decisions; the division of labour between territories within and between countries, leading to steep centre periphery gradients; and the expansionist nature of capitalism in space, controlling more and more of economic cycles, by means of monopolies, in general. The important first characteristic, capital as measure, is also taken up implicitly but since Marxism is relatively weak on alternative measures, Marxists, like liberals, came to behave like book-keepers, calculating costs and benefits in monetary terms, and differed only as to the entries and how they should be calculated - leading to different conclusions.

Where liberalism is actor-oriented, Marxism is structure-oriented, corresponding to the basic development strategies. Where the liberal development school argues in favour of letting individuals loose (the strong towards the top, the less strong being made more mobile horizontally), Marxist schools would argue in favour of basic structural change first. Reversing the above order, first on the programme comes the idea of detaching the periphery from the centre so that it can constitute a centre in its own right, that is, of doing its own

processing of all local factors, raw materials, capital and human beings. This, however, is only a first phase and entirely compatible with local, national capitalism and is to be followed by a second one with internal centre periphery gradients reduced through collectivization of means of production, the use of the social surplus for the satisfaction of fundamental needs and social planning in general. There is then a possible third stage wherein basic changes take place in the division of labour structure inside the production unit. To many people this probably sounds like "politics" and not "development theory" - just as liberal development theory sounds like apologetics and rationalization, not to say gross mystification, to those whose inclination is more Marxist. We prefer to see both of them as theories of development with accompanying practices, and then move on to a more critical appraisal of the Marxist approach.

It should be mentioned at the outset that the bulk of Marxist writing and practice takes the form of criticism of capitalist society and of theoretically-guided action inside those societies. It is as if most of liberal thinking were concerned with a critique of mediaeval and early modern social formation rather than with the building of the post-industrial revolution society in the West. Much of this asymmetry between the two schools of thought is simply due to the circumstance that the liberal school has been relevant for a considerably longer time span in human history, not to mention for vaster portions of the world, whereas the Marxist school as a development theory in the positive sense, concerned with post-capitalist social formations, has only half a century's experience to draw upon, from a limited part of the world, and only from societies that were formerly periphery countries in the world capitalist system and may even now tend to become reintegrated again. There is no experience with what socialism in a society in the capitalist centre might look like. (22)

In our view, the basic criticism of the Marxist theory of development would not be that it represents too radical a break with the present world structure, but that it is not radical enough in its conception, and for that very reason, not in its practice either - with one possible exception. (23) Moreover, the criticism would also be that there is a pronounced tendency to neglect some fundamental insights from the liberal school, indeed raising the question as to whether some type of synthesis of the two might some day become a valid component of a much richer theory of development. (It would be typically Western to believe that such a synthesis, if at all possible, would exhaust the whole universe of development thinking. There will be more about this in the next section.)

The basic point with which Marxism has not come to grips is the fact that post-capitalist formations found in eastern Europe and the Soviet Union, and also in some of the Third World socialist countries (Cuba, Mongolia, perhaps also North Korea) are entirely compatible with Model II social formations. In fact, they can be seen as variations within that formation, but of a State capitalist rather than a private capitalist variety. We would base this on the idea that the four characteristics of capitalism mentioned earlier are still prevalent in these societies: capital as measure, division of labour in terms of decision-making and problem-solving with the top level recruited differently; a high level of mobility of factors from the periphery to national and district centres, as well as of products in the opposite direction; and an element of expansionism although often only within the geopolitical borders set by the State itself.

Thus, individuals are still detachable and substitutable, highly mobile horizontally and vertically, only the advancement criteria differ with proletarian background, formal education and party loyalty being important factors. (24)

This experience seems to indicate that such societies are able in a surprisingly short span of time to satisfy the basic material needs of the masses of their population; that is, to eradicate poverty. This is a major achievement and sets the socialist Third World countries, by and large, against the peripheral capitalist Third World countries in a contrast which is not complementary to the liberal school of development thinking in general. (25) The liberal school solves this problem, decreasingly even to its own satisfaction, by measuring development in terms of averages, such as growth per capita, whereby some very quickly growing or expanding classes, or segments, or sectors of the population of the countries can compensate for a little, zero or even negative growth at the bottom, or even for the majority. Today this circumstance alone would count very significantly in favour of the Marxist approach. (26)

Some may now argue that this is already convincing enough, that the task was to raise the material level of the masses through collectivization of the means of production and not to depart from the fundamental social grammar of Model II society. This may be so, but two problems still remain, two reasons why the practice and the theory may be said not to be radical enough: (1) what happens after the basic material needs have been satisfied? and (2) is Model II society an acceptable developmental goal? These two questions are closely related since there may be some incompatibility between Model II society and some of the less material but, nevertheless basic needs. What these needs are will be discussed at length in the next section, suffice it to say here that they may perhaps be conveniently divided into two classes, one roughly coinciding with "human rights", and another which for lack of a better term we shall refer to as "non-material

needs". Needless to say, all these classes of needs and rights are open-ended and merely suggestive, but equally evident is that anybody's entire body of development thinking and practice is coloured by what the person or the "school" explicitly or implicitly admits into these lists.

And this is where the emphasis on the satisfaction of basic material needs for everybody, the strong point of the Marxist approach, is in time converted into a weak point. It gives food for thought to consider that such basic needs are also satisfied for animals in a well-organized zoological garden. Animals regularly get the needed physiological input in the form of food, air and water and sleep, and output in the form of movement and excretion. They are usually adequately sheltered in cages exhibiting an astounding variety in the more famous zoos. Clothing offers no problem since, as opposed to the "naked ape", they are usually adequately equipped with fur. There are veterinarians to treat their illnesses, usually on a much more regular basis than in the periphery of world capitalism. And in the most advanced zoos for the most intelligent animals such as dolphins there is even some institutionalized arrangement for some type of education.

Thus, societies releasing productive creativity by means of fundamental structural transformation, getting control over the social surplus in favour of the satisfaction of everyone's basic needs (thereby also taking care of distribution) may easily exhibit similarities with zoological gardens unless other developmental goals, usually not associated with zoological gardens are also brought into the picture. The liberal answer, disposable income to permit the individual to plan his own market behaviour and trade offs, is only a very partial one.

This kind of critique, however, would not only apply to Marxist schools of development, but equally well to liberal schools that have progressed on the idea of economic growth per capita to the idea of satisfaction of basic needs for everybody. The entire approach casts the economist in the rôle of the zoo-keeper or game-warden. He becomes the subject, the population becomes the object, the clients - and he satisfies some of his non-material needs (e.g., for activity, creativity, even self-realization) in contributing to the satisfaction of their material needs. No doubt there is a problem here, the problem belongs to the development problematic and has something to do with Model II society regardless of ownership of means of production.

And this of course is where the second critique against Marxism comes in: the cognitive framework fails to encompass new types of class formation, partly because the very term "class" has been given such a narrow interpretation (related to the relations vis-à-vis the means of production) in Marxist thinking. The concern has been with abolition of class formation in that particular sense, and thus has led to a view of development that can be characterized as "development of structures" - using, for instance, as indicator the proportion of the economy which has been collectivized one way or the other.

What could be at the root of class formation in societies that have undergone this capitalist-socialist transition? Foremost and most basically: if the point of departure is a Model I or Model II society, or a mixture of the two (a society with some late feudal and early capitalist characteristics), verticality is already built into most social relations at the social macro and micro level, e.g. in the family, at school, in work relations, etc. There has been a tendency, perhaps in Marxist practice rather than in Marxist thinking, to believe that with horizontalization relative to the ownership of the means of production other social relations will also tend to horizontalize. They do not, and among the explanatory reasons we can mention the following:

According to the general theory somebody has to lead the revolt in order to transform the structure, somebody has to be the entrepreneur, that is, the most advanced elements in the proletariat and certain non-proletariat elements which have internalized the doctrine and developed loyalty to the entrepreneurial organization, the Party. Although this is not said explicitly it is tempting to add that as in liberal thinking there must also be a reward for these entrepreneurs - and that differential reward is already part of the verticality.

The idea that with societies coupled in such a manner that a general horizontalization will follow with a certain automaticity from a basic change where ownership of means of production etc. is concerned presupposes a view of society as strongly and relatively unilinearly coupled, where consequences follow upon the cause like in a game of dominoes. This amounts to a prophecy, and the question becomes what to do when prophecy fails. If the theory is to be saved other circumstances have to be cited to explain the failure: foreign intervention, adverse climatic conditions, saboteurs, or that the initial conditions were not yet satisfied. There is also the possibility of proceeding by tautology: anything in a society that has socialized the means of production is by definition socialist. Our point, however, is that to administer all this an upper class which controls the means of image formation in society as well as the reactions to external and internal, real and imagined enemies is needed. This is also a basic source of verticalization.

Finally, there is the point alluded to above: in order to administer the social surplus and steer it in the direction of fundamental need satisfaction, given the nature and size of a modern State a group of managers, administrators and bureaucrats will tend to emerge, perhaps of a size more than proportionate to the size of the country.

Thus, following the three directions just given, the basis is laid for the crystallization of new layers at the top of society: a group of former proletarians and other revolutionaries from the preceding social formation, a group of more traditional power wielders tied to the machineries of propaganda, police and the military, and a group of managers, economists and bureaucrats. It is well known that the latter two will in time tend to turn against the first group as time proceeds, so that the organization for revolutionary transformation, the Party, will gradually change its composition from being worker-dominated to being largely intelligentsia, or white collar dominated. Since all this also requires a considerable change of personnel, that is a very high level of mobility, the process will in itself facilitate individualization, detachability and mobility of individuals on the basis of achievement, education and loyalty, and the net result is a Model II society with satisfaction of basic material needs and certain repressive features.

It is at this point that the Chinese contribution to development thinking enters, whether classified as inside or outside Marxism. The basic point seems to be that the Chinese do not want the Model II society, but want to proceed to a Model III social formation - in the figure referred to as communal, a term taken from the major institution in China, the People's Communes. This was the fifth stage in the development in the rural part of China after the liberation in 1949 (the other four being land reform, mutual aid institutions, elementary co-operatives and advanced co-operatives). It may be pointed out that the Chinese tradition already carries a very heavy component of collectivism, perhaps specially related to the Buddhist element, something which is absent in the Western civilization, particularly in the Protestant countries. It may also be added that dialectic thinking, the idea of ever-recurring contradictions, that nothing is perfect and worthy of being consolidated by everything has contradictions built into it (the yin-yang principle) must have made it relatively easy for the Chinese to conceive of new contradictions emerging in the system brought about by the transcendence of the contradictions prior to 1949. Consequently, the Chinese had a better point of departure when it comes to the possibility of seeing horizontalization as an open-ended agenda; after the collectivization of means of production what new type of dominance relations may emerge? The answer given in 1966 was clearly, for instance, "in the field of education/professionalism and in the relation between people in general and the Party". The Chinese may have overdone the idea that these elements were foisted upon them by the Soviet penetration, and perhaps not seen sufficiently clearly that they might have developed even, without Soviet presence, and perhaps due to the shortcomings in Marxist thinking indicated above -

shortcomings for which the Chinese tradition already compensates somewhat, particularly when coupled with the structural insight provided by Marxism.

Be that as it may, it is at least clear that the Chinese have seen development in terms of collectivization of the means of production and use of the social surplus, within the context of social planning, for basic needs satisfaction. This was only the first phase (1949-1966), roughly speaking, and the Cultural Revolution cannot be credited with it. What the Chinese started doing during the Cultural Revolution seems to be very closely related to the three reasons for new class formation given above.

First, there was the idea that nobody was "born red" - everybody had somehow to prove himself or herself. This notion is in itself individualist, and achievement oriented - hence the Chinese problem was how to combine this with a collectivist and horizontal society and the mechanisms are well known: rotation of leadership at the local level, cadre schools, "remoulding of the personality" and so on.

Second, there was a certain dedogmatization, expressed in the formula that after "two-into-one" there will always be a "one-into-two" - a highly shortened version of the basic principle of dialectics as something on-going, never-ending. This was then combined with the idea that the liberation was not only necessary, but was a sufficient condition. The Cultural Revolution went one step further, stating that there would be many corresponding processes in the future.

Third, there was the idea of avoiding professionalism and managerialism from the top. This was done through a high level of decentralization of the economy, through commune-ism rather than state-ism, not only in the field of agriculture but also in industrial production since much of this is performed inside the People's Communes. In so doing some of the centre-periphery gradient of the modern State is broken down. There is an emphasis on local self-reliance rather than on getting directives and aid from the centre. Moreover, the Chinese also went further to try to break down the division of labour within the production unit through efforts to merge workers and engineers into one, the famous "worker-engineer" idea, thus narrowing the gap between the problem-solvers and the doers. This was done partly through rotation within the production unit, and partly through the creation of entirely new types of social positions.

Thus, the Chinese contribution is considerable since it goes much deeper in its development practice, constituting a more fundamental attack on capitalism (all four aspects of it, the Chinese also largely refuse to use monetary evaluation of the production process, preferring measures of yield, and conversion into needs-satisfaction).

This, in our view, is more or less the current situation of development thinking and practice.

There are two broad traditions, and within them there are more conservative and more radical varieties. Needless to say, the schools are conditioned by present power relations and social relations in general, and by historical processes: "Tell me which group in which country you belong to, and I shall tell you which development theory you have". But from this it does not follow that to predict the future development of development thinking one should simply try to predict the future of international relations, postulating that "the leading theories are the theories of the leading countries" (and the leading indicators are those that put them on top). There is a dialectic at work here: development theories also influence power relations to some extent. Hence, let us try to indicate some possibilities for further development of the theories of development.

IV. TOWARDS A NEEDS-ORIENTED THEORY OF DEVELOPMENT

What is being said in this section is in no way an alternative to the liberal and Marxist approaches, it is intended rather to be complementary. The basic point is the re-introduction of man as the focus of development, the purpose and measure of the whole "exercise". It may certainly be said that this also underlies the liberal and Marxist approaches, (27) but the point has to be made again and again if one accepts our basic criticism, that liberal thinking tends to become too enthused with the development of things and systems, and Marxism too enthused with the development of structures, in order to develop things and systems after new creative, or productive forces have been released. Hence, in terms of Figure I, the present section is an effort to shift the point of gravity towards the inner circle, and to some extent also towards the outer one, towards Man and Nature. Our major concern, however, will be the former.

To do so, an image of Man is needed, not only empirical man but potential man if the purpose of development is some kind of "personal growth". (28) This puts Man in the centre of the enterprise, and also brings in the whole tradition of philosophical, not to mention theological thinking against which the social sciences were to some extent a reaction, even an over-reaction since development theory became too tied to the development of society in the sense of things plus systems plus structures. (29)

One way of creating an image of Man is through an image of his needs. This immediately brings in a vision of short or long lists of needs, (30) possibly divided into material and non-material, basic and less basic - and what shall be done here is in that tradition.

One shortcoming of the tradition should immediately be pointed out: it is analytical, fragmentary, chopping Man into a set of needs rather than trying to develop a holistic conception of Man. In so doing it is certainly in the Western analytical tradition rather than in the Eastern more comprehensive, perhaps intuitive tradition of understanding Man, a shortcoming readily admitted. It is also well known what this easily leads to: an institution-building around each need, both at national and international levels (ministries of health nationally, a world health organization internationally, both built around the need), converting analytical fragmentation into institutional fragmentation, with well-known harmful consequences. This is also clearly related to the size of the political units in our period of history, the nation-States and the quest for some kind of global institutionalization, and is related to analytical abstractionism, the emergence of intellectuals and professionals to handle human affairs already referred to in the first section. As one of the non-material, but basic needs we are going to emphasize, is the need to be active, to be subject and to be autonomous. This process is counter-productive, since the only place where more holistic conceptions and processes aiming at needs satisfaction can emerge seems to be at a more local level, in a "decentralized" setting, and perhaps even in non-Western cultures, at present, or less male-dominated societies.

We have mentioned all this by way of introduction since we do not believe any more in a development process that fragments Man according to needs in a set of global institutions, some type of world government, than in the current fragmentation of Man into geopolitical units known as States. There are other possibilities, but it may very well be that those are seen most clearly if we try to enrich our conceptualization of Man, and that would be the answer so far to our own critique.

The next problem we have to tackle, again by way of introduction, is that of universality versus geographical and historical specificity of needs. The difficulty is clear enough. On the one hand, there is a need to get a rich concept of Man, with many connotations, here translated into relatively comprehensive lists of needs. On the other hand, the variations in time(31) and space are tremendous and any imposition of such lists from anywhere as directives for development processes would constitute some kind of neo-imperialism. It is also obvious that one way in which this contradiction has been overcome in the past and currently is through short lists, focusing on the basic material needs so as to obtain something close to a universal consensus. The difficulties with such a consensus have been spelt out in the preceding section. It does not constitute a solution either.

In fact, we do not think there is any solution in the sense of a stable, universally agreed-upon list. The list of needs is in itself a part of the development process, currently needing revision as the general process goes on, and sensitive to all kinds of differences in space. How this revision is carried out, by experts or by the people themselves - according to Model I, Model II or Model III social formations is also a part of the developmental process. But one approach which might be interesting would be to think in terms of maximum lists and not minimum ones, with the idea that these lists constitute a frame of reference, not to be taken literally in all details, but to be looked into when a development process is to be initiated or evaluated. Since a minimum list provides insufficient guidance and a maximum list may lead to oversteering, to frustrations due to incompatibilities and to unfortunate trade-offs, what we have argued for, is that some kind of flexible, indicative long list might be more suitable.

The problem, then, obviously is what to include in such lists. What are the criteria that something has to satisfy in order to be classified as a human need? And what is the meta-criterion according to which such criteria are selected.(32) The answer to this is at a higher level of abstraction, but we will try it none the less. In our view the conditio sine qua non criterion (the idea of a necessary condition) would be some kind of "system maintenance"; initially of the human body, and secondly of the society itself.

In other words, we could conceive of the following two criteria to designate something as a need:

(1) If it is a necessary condition for a human being to exist, it is a need. In other words, if non-satisfaction leads to the disintegration, destruction, non-existence of the human being.

(2) If it is a necessary condition for a society to exist over longer time, then it is a need. In other words, if the non-satisfaction leads to disruption, disintegration, or non-existence of the society, for instance, through revolt or non-participation, apathy, anomie.

This means that we are operating at two levels where the criteria are concerned, but both are fundamentally linked to human beings as such, not to the things-systems-structure abstraction. At the first level, one would simply be concerned with the existence of the human being as such. At the second level, one would try to derive the needs from observations of what human beings fight for, what makes them withdraw and try to disrupt the social "order" in which they are embedded. There is no doubt that this criterion is problematic and some of the problems will be discussed later on, but it is better than no criterion at all, at least at the outset. Take a need, or a class of needs, such as that for freedom. However it is conceived of, it is a fact that Man continues to exist even when he is deprived of freedom in very many senses of that term. But it is equally well known that he sometimes stands up and fights in order to achieve it.

Both criteria mentioned are empirical in the sense that they give us some possibility of deriving needs from empirical observations. Being empirical they also have the advantage of being sensitive to the variations in space and time. Human beings and social orders may persist or succumb under different conditions depending on where in space and when in time they are located. Physical disintegration may be brought about by different conditions depending on where one is located; what people stand up and fight for may vary very much through history.

These criteria also aim at solving another problem: limiting the freedom of experts to impose their own conceptions of needs. As formulated, the burden of proof lies in Man's behaviour, not in élite postulation. On the other hand, as already mentioned: élites rather than people in general think in terms of such lists, although we are relatively convinced that they are closer to the way in which people, when left to themselves would think about development than most economic indicator lists developed by specialists in abstract economics.

The following, then, is an effort to establish such a flexible maxi-list. As the reader will see it has many elements, and these are divided into four categories, in the usual fashion: from the more to the less fundamental. This means that needs at a lower level have to be satisfied <u>at least to some extent</u> for need-satisfaction at a <u>higher level</u> to take place. (33) One has to be alive for feeding to be meaningful,(34) one has to be fed for politics to be meaningful; some kind of politics are needed for the last ten needs, etc.

Table I

VALUES: BASIC NEEDS, MATERIAL AND NON-MATERIAL

	Category	Needs and/or rights	Goods/services
SECURITY	Security	(Individual: against accident, homicide (Collective: against attack, war	SECURITY
WELFARE	Physiological	(Input : nutrition, air, water, sleep (Output : movement, excretion	FOOD, WATER
	Ecological	(Climatic: protection, privacy (Somatic : protection against disease, health	CLOTHES, SHELTER MEDICAL TREATMENT
	Socio-cultural	Culture: self-expression, dialogue, education	SCHOOLING
FREEDOM	Mobility	(Right to travel and be travelled to (Rights to expression and impression	TRANSPORTATION COMMUNICATION
	Politics	(Rights of consciousness-formation (Rights of mobilization (Rights of confrontation	MEETINGS, MEDIA PARTIES ELECTIONS
	Legal	Rights of due process of law	COURTS, etc.
	Work	Right to work	JOBS
	Choice	(Right to choose occupation (Right to choose spouse (Right to choose place to live	
IDENTITY	Relation to self (individual needs)	(Need of self-expression, praxis, (creativity	HOBBIES, LEISURE
		(Need for self-actuation, realizing (potentials syn- and diachronically	LEISURE, VACATION
		(Need for well-being, happiness, joy	VACATION
		(Need for a sense of purpose, (a sense of meaning with life	RELIGION, IDEOLOGY
	Relation to others (collective needs)	(Need for affection, love, sex, (spouse, offspring	PRIMARY GROUPS
		(Need for roots, belongingness, support, (association with similar humans	SECONDARY GROUPS

Table I (Cont.)

	Category	Needs and/or rights	Goods/services
IDENTITY	Relation to society (social needs)	(Need to be active, to be subject, (not passive, client, object	
		(Need to understand what conditions one's (life, for social transparence	
		(Need for challenge, new experience - (also intellectual and aesthetic	
	Relation to nature	(Need of some kind of partnership with (nature	

A few words of comment may be in order.

First, this list to bring in the human rights tradition, in addition to that of human needs, thus bringing into the field of development not only biologists/physiologists (already present in connexion with nutrition), but also psychologists, social psychologists, and lawyers. There is a purpose in this. If development concerns all men and all of man, it should at least concern all the social sciences and natural sciences and as much of the humanities as we are able to include. It cannot possibly be the concern of one discipline only. And, of course, the human rights tradition belongs: it has built into it an image of the good society, as well as so many other images. Indicators of development, hence, should also include indicators of the human rights situation in a country. But having said that it should also be added that the human rights tradition has had a tendency to ignore the more fundamental needs, perhaps relegating them to more residual categories like "economic and social rights". (35) They are more concerned with the rights of élites in liberal, Model II societies than with those of people in general everywhere - a point we shall not elaborate further here. (36)

Second, we have added, at the right, some indication of the "services" or "goods" (sometimes material, sometimes institutional) that have to be provided one way or another in order for the needs to be satisfied. Some of them are put in parentheses because the relation is far from obvious, as in the case where mention is meaningful only with a special socio-cultural context, e.g., rich Western countries. The list, nevertheless, gives some indication of what has to be produced, as a minimum, for need satisfaction at a basic level to take place. For the last needs on the list goods/services are less institutionalized, if at all available - they may even be counteracted.

Third, we have started with the "most fundamental right", the right to life, to survival - pointing to such obvious factors as the need for including the probability of dying in an accident, a war or at a murderer's hand in the concept of development. (37)

Then, we proceed to the conditions, given a

good chance of survival, under which human beings emerge. Some of them are physiological, some have to do with protection against the natural environment, and some are social. If the latter category is not satisfied there may be biological survival of the individual, but togetherness and self-expression, including dialogue - based on mastery of some language, not necessarily only verbal - is a condition for human beings to grow. We have doubts as to the need for children, that it is perhaps more a societal need than a human one (and perhaps even an élite need, it is usually élites who argue in favour of more population) - but some form of togetherness is postulated as a need, and in terms of nearness suggested by such words as love (not necessarily Western, "romantic" love - may be that was a part of the transition from Model I to Model II society, individual choice and mobility), and sex.

Then we proceed to the human rights knowing that the list could be made longer. Like the foregoing list, it contains the kinds of things that people stand up and fight for, not necessarily for the particular institutions which have emerged around rights in Western societies, but for the basic rights. (38) It should be pointed out that in saying this, politics, for instance, is not seen merely as an instrument for wise decisions and social cohesion, but as something important for personal growth, i.e., as a form of self-assertion and self-expression, of the individual self as well as the collective self.

Finally, there is a list of ten additional less tangible needs. A distinction has been made between relations to self, to others, to society and to nature - and many of the elements in the list are taken from the maslowian tradition, which again is based to some extent on non-Western cultures. (39) The list is certainly much less definitive, which is no shortcoming according to what has been said above, for two obvious reasons. Firstly, these are the kinds of needs that are not well satisfied in societies which are based, largely on vertical division of labour resulting in alienation and sometimes also in exploitation in the sense of repression or maintaining people below any acceptable poverty line.

Secondly, since these are the societies that set the goals of development and because they are Centre, Western societies, such needs are usually seen as "intangible", "non-measurable", "philosophical", etc. In addition, there is less consensus about them because it is not obvious that the two criteria mentioned earlier in this section are satisfied. Human beings continue to exist even with little new experience, and they do not stand up and fight for it, either. However, we also included the criterion of non-participation, and we have also mentioned apathy. Less socio-culturally biased (by the conditions of Model II societies) research might have much insight to add on this.

By and large, it follows from what has been said in the two preceding sections that whereas the problem of direct violence may affect all social formations, the problem of poverty is particularly important in the periphery of world capitalism (the "Third World" countries), the problem of human rights particularly important in those societies as well as in some of the socialist countries, and the problem of the more advanced needs is perhaps most important in industrial societies in general. Perhaps these are exactly the types of insights that richer concepts of development and better indicators might make it possible to explore systematically.

So much for the needs and their satisfaction, i.e., the goal, the purpose of development. What about the instruments, the means to achieve these ends? A development theory is not only a theory which accounts empirically and theoretically for the present and projects an image of the future; it also has to say something about the strategies.

However, this question throws us back to the very point of departure in this entire exercise. As long as the unilinear tracks of development were accepted the answer to the question was simple. The image of the future did not have to be described in value terms, but one could just read off empirically some basic characteristics of the goal state countries, i.e., the developed countries, particularly the United States and/or the Soviet Union. The strategies were also relatively clear: imitate these countries as far as possible. In short, development theory and practice became largely an empirical discipline studying very difficult problems of applicability and transfer, but fundamentally tying development studies to the empiricist, even positivist tradition in the social sciences. The critique of unilinear perspectives on development is at the same time a critique of that tradition, not rejecting it, but seeing it as incomplete.

A needs-oriented theory of development leaves the field more open since we know very little about under what class of conditions a large sub-set from this needs list would be satisfied. Concretely, this means that every society, big or small, would and should feel free to formulate its own policy of development, to build the path towards a higher level of satisfaction of needs. (40) A theory of development could indicate some steps along that path. Since

the points of departure are rather different around the world, it might be easy to think in terms of conditions, "steps", close to the achievement of the goal, and particularly in terms of necessary conditions. These conditions, then, would be located in the things-systems-structure circles, and the question becomes: what is the minimum to be said about these three, as well as regards Nature in a theory of development?

Minimally, as far as things are concerned there certainly has to be some socio-economic production, or socio-economic growth for that matter, for material needs to be satisfied - taking "production" in a material sense. Priority should be given to food, clothes, shelter, health hardware and education hardware: we might also add transportation/communication hardware as a condition for freedom in our period of history, so heavily on mobility and other services.

The minimum that can be said about systems, the distributive aspect of development, would be that there must be a high-level diversity, within the society, for the simple reason - derived from general ecological thinking - that if societies gamble on only one approach (e.g. for the solution of health or educational problems), then vulnerability increases. When contradictions pile up within the system the society may even be paralysed, whereas gambling on a variety of systems makes it possible to benefit from a variety of approaches and to let the contradictions unfold themselves in all of them, thereby leading to much richer societies.

In addition, there is the problem of equality and social justice as defined above. The assumption behind need-satisfaction is that it is for all, initially, regardless of ascribed characteristics, secondly, if not exactly at the same level, at least without too much disparity. In fact, both from a liberal and from a Marxist point of view one might today argue that the development process starts at the bottom, with those most in need - thereby building higher levels of social justice and equality into the process from the very beginning.

These four points: socio-economic production, diversity, equality and social justice are held to be necessary conditions. Their inclusion in the development theory is of a tautological nature. Nevertheless they should not be confused with the goals of development. Even if all four criteria are met in a given society, it may very well be that something has gone wrong somewhere in such a way that the purpose of development is still not met. They are not sufficient conditions.

But what is the minimum that can be said about structures? This is more problematic. If it had not been, there would have been no liberal-Marxist controversy. The reader will have already understood that whereas what has been said so far (goals, things, systems) is, if not within, at least, compatible with a liberal tradition, what follows would be more in the Marxist tradition. By and large, we would feel that the Marxist structural analysis is correct although it has pointed to necessary rather than to sufficient conditions, and is too limited in its economism and its determinism, at least in some versions of Marxism.

In general, we would feel that the necessary condition for development to take place is a reduction, if not outright abolition of centre-periphery gradients. These gradients seem to have at least four aspects built into them: exploitation in the sense of vertical division of labour between centre and periphery (economic division of labour brought about by capitalism, political division of labour by State formation with the two going hand in hand); penetration characterized by the centre framing and moulding the mind and the action of the periphery, partly through local bridgeheads (the centre in the periphery); fragmentation of the periphery (Third World countries, districts inside countries, or sectors of the economy that are not linked to each other locally but tied to well-integrated sectors in the centre economies - Samir Amin's important thesis) and marginalization, the definition of internal and external peripheries as second class, only partly participant in the development process, mainly being kept as a reserve (e.g., the reserve army of workers under capitalism).

The antonyms of these four structural formations, and hence the developmental instruments, would be equity (or horizontal division of labour), autonomy (or autarchy which now tends to be the most important economic expression of that idea), solidarity (e.g. in the form of trade unions of exploited workers, exploited districts, exploited countries in order to improve their terms of exchange and terms of trade, and in order to change the structure in general), and participation of everybody on an equal footing, people, districts and countries. All this is rather antithetical to capitalism as described above (particularly the last three of the four characteristics). The acceptance of equity, autonomy, solidarity and participation as developmental instruments is antithetical to at least large-scale capitalism, private or State, as we know it today. It may not, however, be antithetical to small-scale local capitalism on a self-reliant, community basis, but then it may be argued that such an animal does not exist, that it is in the nature of capitalism to try to grow big, initially filling the space provided for it geopolitically by the nation States; secondly, feeling itself strait-jacketed and engaged in spillovers of an international nature. (41)

However, the structural dimensions made use of here are by no means limited to an analysis of economic relations. There is also vertical division of labour between the leaders and the led, between rulers and ruled, between teachers and pupils, between senders and receivers of communications and so on. In the work of the present authors, the four dimensions of exploitation, penetration, fragmentation and marginalization constitute the mechanisms of structural violence, and when operating transnationally, the structural mechanisms of imperialism, whether that imperialism is primarily economic, military, political, cultural, social or communicative. (42) Particularly important in the today world is social imperialism, a term much employed by the Chinese, which might be conceived of as the imposition by one country of its own vertical structure on another, using those at the top of the local vertical structure as bridgeheads for some of their own designs - not necessarily for economic benefits. (43)

The verticality of this relationship is the verticality between the centre of a total social configuration, for instance, a model of post-capitalist society, and the receiver of that model. The imperialist nature (implicit in the word "imposition" used above) shows up when the periphery tries to change this structure. The extension of the concept of structural violence and structural imperialism to non-economic fields automatically sensitizes the researcher to types of exploitation other than economic ones, and expands the concept of development from essentially an effort to become less poor, through an effort to become less dependent economically to one to become less dependent in general, to become autonomous, self-reliant, and the master of one's own social dialectic not the consequence of a cause located in one or more outside centre countries.

Finally, there is the outer circle - Nature. It obviously has to enter in any seriously designed and practised theory of development. If the goal and purpose of development is the development of Man, then production of things, at least some things, is one necessary condition, with another being the distribution of these things. (44) Then there are the constraints; the man-made constraints in the form of domestic and global structures just mentioned, and the non-man-made, natural ones, today customarily referred to under the heading of the finiteness of Nature. Instead of that one might perhaps rather approach the problem in terms of a theory of ecological balance, Man being a part, not a master of that generalized concept of ecology. When the ecological balance breaks down, it shows up, among other things, in the form of depletion of resources and pollution of Nature and Man alike. This is not a new phenomenon in human history, but current expansionist production processes based on economic cycles are so world-encompassing that spontaneous control processes which should operate at the more local level (because the producer will sense the impact of his depletion, and producer and consumer alike will sense the impact of this pollution) disintegrate leading to many ecological breakdowns. Hence, a basic aspect of the entire development process is to what extent this ecological balance is maintained, not only on a general global level, or a regional or national level, but also on local levels - particularly if it is felt that developmental goals can be best promoted by stimulating local self-reliance. (45)

This type of thinking has introduced a new and extremely important dimension in development theory and practice: the concept of a maximum, a ceiling on development - for instance as exemplified in the title of the first study of the Club of Rome, Limits to Growth. Production is needed in order to provide a minimum material need-satisfaction, a floor level below which nobody should be located when it comes to consumption of food, medical care, schooling, transportation and communication. But it is also true that there is a limit to how much can be produced given the finiteness of nature and the vulnerability of ecological balances and, since Man is a part of Nature, the increasing size of the human population. Maybe one day that maximum can even be calculated, although in a dynamic manner, since new knowledge will make it a non-static concept. The question then becomes where the maximum is located relative to the minimum and whether there is a positive window or a negative window with the maximum below the minimum - indicative of impending catastrophe. So-called experts differ on this (our feeling is that the window will be positive for some time to come), but it does raise the question of whether one should not think in terms of social maxima already now, i.e. of limits to the consumption of food, clothes and shelter, etc. by the rich, not only conceiving of development in terms of how to raise the floor level for the poor.

There are, of course, also two other reasons for arriving at this conclusion. The rate of consumption in the rich countries (and the rich pockets in the poor countries) threatens not only the limitations set by Nature, but also those imposed by Man - both have a limited capacity. Too much food, clothes, shelter, medical care, schooling, transportation, communication, etc., obviously become counter-productive at the individual level. At the same time too much disparity in terms of what people have, and what countries have on the average, also constitutes an important power differential. Inequality may be converted to abuse of power and, in turn, to the creation of a new centre-periphery gradient. To some extent this brings new perspectives on the old debate on equality with equality being conceived of not as a point, but as a range of variation between social minima and social maxima that should be set not by experts but, somehow, by the population in general. Needless to say, we are not anywhere near any kind of institutionalization in that regard.

To conclude: we have tried to present an image of a relatively rich development theory, perhaps drawing on Western thinking in liberal and Marxist varieties as to how to conceive society, the production of things, the distribution of things inside systems, the transformation of structures; perhaps, implicitly rather than explicitly, drawing on more Eastern thinking as to how to conceive of Man, particularly "inner man" (the last ten in the list of needs, for instance), and how to conceive Man's relation to Nature. At the same time, the theory argues for broader thinking on development, not only in terms of underdevelopment where social minima are not satisfied, but also in those of over-development where social maxima are overstepped. Thus, there may be said to be two forms of development, an underdevelopment where too little is produced to satisfy the needs, and an over-development where too much is produced and in such a way that some needs are over-satisfied while others, particularly the more ephemeral ones towards the end of the list, are left entirely unsatisfied. And with that conclusion, the basis exists for the idea that all societies are mal-developed, only in different ways; that we are all parts of the same global dialectic, and that the time has come to transcend the centre-periphery idea implicit in the "developed/developing" dichotomy, in favour of the idea that we are all mal-developed one way or the other.

Appendix

TOWARDS NEW INDICATORS
OF DEVELOPMENT

The changes in development theory and practice currently taking place have to be accompanied by corresponding changes in the indicators of development. The conventional indicators, GNP per capita and related measures, only served the purpose as long as development was identified with economic growth, and the latter was above all identified with processing and trading. If development is to be identified with such components as:
- satisfaction of human needs for all,
- equality and social justice,
- level of autonomy, or self-reliance, with participation of all, and
- ecological balance,
then development indicators will have to reflect exactly this, as directly as possible.

In addition to being misleading as an indicator, the GNP per capita tradition also suffered from some other serious shortcomings. Thus, it was a measure developed by experts for experts, with no popular involvement whatsoever, and based on a calculation process beyond the comprehension of most people. An indicator supposed to reflect something as basic as the level of development should at least be easy to calculate and understandable for everybody with a minimum of education, in other words, for those most concerned. It should not be permitted to constitute part of a class barrier between those who understand and those who do not. Ideally, it should also be the object of continuing debate and reappraisal, and not only among experts. Indicators of development should be at least as dynamic as the development process itself and not like the GNP (per capita) remain basically unchanged in spite of tremendous changes in the theory and practice of development.

Development, however it may be conceived, is a complex process and will have to be reflected through a set of indicators rather than by means of one indicator. To arrive at such a set, there are at least two lines to pursue by defining either:

(a) areas of human needs, or
(b) aspects of satisfaction of human needs.

The following are some suggestions concerning these lines of thinking about development indicators.

Some areas of human needs (not necessarily in order of priority) that probably will be on most of such lists are:
- food and water
- shelter and clothing
- health
- education.

These are usually referred to as basic or fundamental needs, but the development process does not end with their satisfaction. There are other needs, such as:
- work
- freedom of publication and expression of ideas
- freedom of movement of persons
- politics

and many others could be mentioned (e. g. togetherness, friendship, love; need for respect; need for joy and to be a source of joy for others; happiness; self-realization, the need for a sense of meaning of life). Incidentally, no reference will be made to such conventional distinctions as between economic, political and social indicators, for human beings do not live in compartments that reflect unfortunate division of labour among social scientists more than anything else.

Both lists are problematic. For each item there is usually a narrow, sometimes even perverted, concept that only very partially reflects a much richer, more deeply human, underlying concept.

Thus - can food be reduced to proteins and calories etc., or is there also a quality to food, even as an act of social communication?

Can shelter be reduced to square metres of covered space per person, or does one have to take into consideration the entire habitat of the individual?

Can health be reduced to longevity and access to medical care, or does one have to take into account quality of living and capacity for curing oneself and others?

Can education be reduced to schooling, the number of years (and levels) spent at school, or does one have to take into account capacity for critical and constructive dialogue, for understanding the human and non-human environment

so as to engage in development process together with others?

Can work be reduced to alienating jobs, and employment, or does one have to take into account the level of challenge and opportunity for creativity and self-expression built into the work? And what about time lost in commuting to jobs far away from home, should that not somehow enter into the picture?

Can freedom to publish and express be reduced to access to mass media and communication in general, or does one also have to take into account the quality, the truth in a broad sense of the culture being communicated? At any rate, it has to include cultural creativity, not only consumption.

Can freedom of movement be reduced to access to transportation, which may simply lead to moving around, according to job requirements, or does one have to take into account quality of experience?

Can politics be reduced to parliamentarism, or is it a deeper process involving consciousness-formation - which in turn would be based on education and communication - mobilization - which probably would have to be based on communication and transportation - and some element of confrontation and struggle, hedged around by certain rules?

The idea is not to reject the more simplistic answers for proteins, calories, square metres per person, longevity, level of schooling, level of employment, access to communication, access to transportation and some institutionalization of political struggle are all important. They should, however, be seen as approaches rather than as the answers to the problems raised by the more general need for formulation. Behind the whole idea of diversity of development lies the idea of several roads to satisfaction of human needs, not only those most commonly found in national and international statistics.

Among the aspects of satisfaction of human needs current development theory and practice would lead one to focus on:
- level of satisfaction
- distribution of satisfaction
- structure of satisfaction
- ecology of satisfaction.

For each area of human needs all these aspects (that correspond to the four components mentioned under (1) above) should be considered. The following are some suggestions as to how this might be done.

As to the level of need-satisfaction, the conventional method of calculating averages should be abandoned in favour of simply finding the percentage of the population above and agreed upon social minimum. This social minimum should not be identified with a poverty or subsistence line: one should aim higher than that. Needless to say, such minima would vary from place to place and over time, as well as with age, sex, and other characteristics. To impose a standard universal criterion would constitute violence. Using this method one avoids the old pitfall of having the overfed compensate statistically for the underfed, the oversheltered for the undersheltered, etc. Also, the concept and the procedure are very easily communicated; but the data are not easily available - for obvious reasons: they are too politically revealing. Incidentally, it should be pointed out that the concept of satisfaction used here is "objective" relative to a social minimum, and does not necessarily reflect the subjective perception of satisfaction.

Regarding distribution of need-satisfaction, the conventional method of arriving at equality indicators by calculating variances and gini-indices should be abandoned both because they are too complicated and do not give enough information about the absolute distance between high and low in society. Instead, one might simply find the percentage of the population above a social minimum mentioned above and below a well-defined (if not necessarily agreed upon) social maximum. Together with information on the location of minima and maxima this would give a good idea about the social distribution of, for instance, housing, access to medical care and access to schooling and the rest. Thus, the idea of equality would not be that everybody has exactly the same, but so that the range of variation between floor and ceiling, or minimum and maximum is limited.

There is another aspect to distribution: social justice. One way of conceiving of social justice is simply as follows: the level of need-satisfaction should not depend on sex, race, age, whether one lives in a city, town or village, whether one is employer or employee, who one's parents were etc. The yardstick is again very simple: find the percentage of men above minimum in, say, schooling, and the percentage of women, and compare the two percentages. If there is social justice they should be equal, which may mean that they are equally under-schooled, or adequately schooled - and may be compatible with equality or inequality. Social justice does not reflect level of satisfaction or equality in the population, but is nevertheless a very important dimension because it may express the degree of racism, sexism and "ageism" (discrimination against the young and the old by the middle-aged) in the population. What equality is to the difference between individuals, social justice is to the difference between groups.

In dealing with the structure of need-satisfaction, indicators in this field are more problematic, but in general they should reflect the level of self-reliance. Depending on the area the unit of self-reliance will sometimes be local, sometimes national, and sometimes regional, or sub-regional. In some cases it will also make very good sense to talk about the individual human being as the unit of self-reliance: i.e., in connexion with the capacity for self-cure and for self-education. In general, the question to be asked would be: to what extent is the unit capable of being self-sufficient, meaning capable of being its own producer, of what is needed for food, housing, health and educational purposes, work creation, culture, mobility and politics, and to what extent does it depend on other units? Essentially this would be a measure of the extent to which a part of the world is a periphery dependent on some centre elsewhere: a centre dependent on a periphery or a centre in its own right. Many such measurements exist already; some in terms of institutional location, others in the form of budgetary allocation. In the field of mass media, translations, etc., information is available about how much of the culture and ideas communicated are produced locally.

It should be noted that the idea of self-reliance is not that of self-sufficiency but capability of self-sufficiency - reliance on one's own forces so that in a crisis, an emergency, one is in fact self-sufficient. In ordinary periods, self-reliance does not exclude trade and exchange in general, but it does exclude the dependency on such exchanges that would make the unit vulnerable to blackmail. It should also be noted that the antonym of self-reliance is not only dependence on a centre, but also exploitation of a periphery. A self-reliant unit neither exploits, nor is it exploited.

Another structural aspect is mass participation. Development should be for the people, and the extent to which this is true is measured by the aforementioned indicators. But it also should be by the people, and for this to happen, national or even local self-reliance is only a necessary, not sufficient conditions for it may still be in the hands of local owners of means of production, of knowledge, or of expertise. Participation means that not only decision-making, but also production processes are organized in such a way that everybody has a say; not only at meetings where decisions are taken; not only over conditions of the production process but over the technology and the means of production in general. The antonym of participation is not only centralized and/or autocratic rule, but also the rule by experts and professionals in general. For the time being no known indicator exists in this important field.

Moving on to the ecology of need-satisfaction, the key here is how well our "only one world" will be able to sustain not only the present generation, but also reasonably-sized future ones. Indicators would reflect, within the level of the unit of self-reliance, generally not the whole world, the extent to which the processes generating renewable resources are intact, and non-renewable resources are either left untouched, or adequate renewable substitutes are found. Sooner or later, thinking about such indicators will probably have to approach the extremely difficult problem of what would constitute minimum, optimum and maximum population for the various units of self-reliance.

Participation in indicator-formation. What has been said above is nothing more than a sketch of possibilities that would bring indicators more in line with new trends in the theory and practice of development. According to the ideas of self-reliance and mass participation ideally, one should, and indeed might, have mass discussions all over the world on at least the following themes that are basic to the whole concept:

- What are, in fact, the "human needs"?
- What are the priorities in case of conflict?
- What are the trade-offs?
- Where are the cutting points?
  for the floor, the social minimum?
  for the ceiling, the social maximum?
- What would constitute a reasonable level of equality, of social justice - along which dimensions?
- Which are the units of self-reliance for the various human needs?
- What is the meaning of participation?
- What are the minimum and maximum limits for our responsibility to future generation?

Again, the answers will vary profoundly from place to place and over time; this would serve as a safeguard against the use of need-based indicators to standardize, to make uniform patterns of development - for experts, being trained the same way, are usually more similar to each other than to people in general. People who establish indicators should, therefore, be prepared to calculate several versions, depending on variable cut-off points and "units of accounting" (which in this context is what the units of self-reliance are).

The need for new statistics. One basic point about the indicators suggested above would be that they are goal-oriented in the sense that they measure the attainment level of developmental goals. Thus, there is no indicator for "industrialization" or "urbanization", because it is not clear that these constitute developmental goals. At most, they are means, and may not even be that. To assess how a country, or a district within a country or a region of countries is doing, data is needed, and should be presented so as to reflect directly how many (or few) are above which minimum; how many are within which minimum-maximum range; what is the level of social justice; what is the level of self-reliance and how are the ecological parameters. The United Nations, her agencies and the statistical bureaux of Member States should be called upon to collect and present data in a way that clearly reflects new thinking about development. If this is considered too expensive, sample surveys should be used, and/or ways and means should be found whereby people themselves might collect and present the data.

At any rate, more attention should be given to the definition of development indicators. In the past, the leading indicators were the ones that ranked the leading nations highest, such as those on top of the international division of labour (which is what GNP really measures). In the future, the indicators should measure the level of living for the common man and woman everywhere, in a social setting, and mindful of the outer limits set by our finite nature - to serve as a mirror in which people can see themselves and judge how the society is doing locally, nationally, collectively, relative to itself some time ago and relative to other societies when that is appropriate. And here we would even reintroduce the idea of growth; not as the accumulation of machines, of production and of marketable goods, but in the sense of the dynamism of progress, along the many ways to true development: the development of human beings.

NOTES

This paper has been inspired by discussions of many groups.

First, there was the Working Group on Human Resources Indicators convened by the (then) Methods and Analysis Division of the Department of Social Sciences, Unesco, which led to a number of meetings during the years from 1967 to 1973, and to some extent continued by a series of Expert Meetings on Indicators of Social and Economic Change. The present study was commissioned by Serge Fanchette who, like Erwin Solomon and Ramalinga Iyer, has provided the participants at these meetings with very stimulating input and ideas.

Second, there was the World Indicators Programme at the Chair in Conflict and Peace Research, University of Oslo, of which this paper is an outcome. Some of the ideas are a continuation of those developed in the basic programme paper of that project, Measuring World Development (published in the journal Alternatives, Vol. 1, 1975, Nos. 1 and 4).

Third, there were the many discussions at the courses of the Inter-University Centre of Postgraduate Studies, Dubrovnik - particularly the course World Future Models that took place in January 1975.

Fourth, there were the stimulating discussions at the Institut d'études du développement, Geneva. And finally, the conference organized by the Government of Algeria and the International Development Centre in Paris - in Algiers, 24-27 June 1975 - when the Appendix to this paper was presented for the first time.

The basic problem to be explored in this paper, posed by Fanchette, was also suggested by Michel Debeauvais: can we have indicators without a theory of social change? Our answer - as will be seen from the text - is "yes, but not without a theory of human fulfilment", given here as the elements of a theory of human needs. If development is to be defined as development of human beings, then it is at the level of human beings that development has to be measured - if it has to be measured.

(1) A good example here would be Venezuela. Thus, the Venezuelan Government's income from oil increased from Bol. 8 billion in 1972 to Bol. 13 billion in 1973 and to Bol. 45 billion (after the OPEC action) in 1974, while the percentage of the population gainfully employed in agriculture decreased; the percentage employed in the oil industry and in mining also decreased (down to 1.3 per cent in 1974 from 2.8 per cent in 1950 presumably due to high productivity); the percentage in the tertiary industries increased (to 48 per cent in 1974 from 37.8 per cent in 1950); with the percenage of unemployed increasing (from 6.3 per cent to 7.3 per cent). Data from Latin American and Venezuelan sources, in special issue of Kontakt, No. 6, 1975-1976, Copenhagen.

(2) The case here is, of course, the People's Republic of China - seemingly ranking very high on non-material values in the "identity" cluster (see Section 4 in the text), but low in the "freedom" cluster.

(3) In the history of philosophy there seems to be a periodical "rediscovery of man" as goal. The authors of this paper are of the opinion that the indicator movement in general will have much more to desire from philosophy than from the social sciences, once more an indication of the inter- and transdisciplinary character of the indicator movement.

Aristotle in his Politica gives examples of how the politicians of his time were only aiming at what was "most useful" and profitable. They were also confusing the means - economic welfare - with the goal: human well-being (Works of Aristotle, II, 9 1269 a 34-35).

For a very recent formulation of the Protagoras homo mensura principle, see the Cocoyoc declaration - circulated as a document of the General Assembly under the symbol A/C.2/292.

(4) Today this problem is discussed in many fora, sometimes with other answers than "human beings" to the question of what is the goal. In Norway, for instance, the philosopher Arne Naess argues in his book Økologi, samfunn og livsstil (Oslo, 1974) that nature as a whole should be the goal for all activity, and in this general picture man plays an important but not superior rôle. When we in this paper put human beings in the centre, it is for many reasons, and we shall here underline two of them. First, we want to emphasize a positive view of humanity, we see human beings as basically "good". Implicit in this is the belief that the human being, permitted to be human, also will develop in a way that does not hurt nature. Second, we feel that placing man in the centre opens for a much more democratic approach since most people have some kind of view of themselves and their needs and how

they can best be satisfied, whether their consciousness is "false" or not. This feeling in all human beings serves as a point of departure, not dogmas from the point of view of economic and/or social theory. More problematic, however, is how to characterize the ideal state of nature, because, among other reasons, we cannot consult nature and therefore would be inclined to impose upon nature our own conception of what nature should be like. Hence, anchoring developmental concepts in development of nature seems problematic. In a sense we are more in the tradition of Durkheim when he placed as a primary goal the maximum cohesion and conviviality between members of society, a higher goal than economic growth and development in general. This was one reason why he rejected both liberalism and Marxism (reform socialism). In his opinion both of them had material growth as the ultimate goal, and differed only as to the means.

(5) The term is used by the senior author in The True Worlds: A Transnational Perspective, New York, 1977, and is the autotelic value in the World Indicators Programme at the University of Oslo.

(6) "Damage" stands for a number of things in a theory of needs. We are thinking of the discussion concerning the hierarchy of needs, and the idea that some needs are more fundamental than others. Among the criteria that are suggested in the literature of needs one might mention:

(a) necessary conditions for survival in a purely physiological sense;

(b) necessary conditions for preservation of mental health - as this is explored by Maslow, Motivation and Personality, New York, Harper & Row, 1970;

(c) "Basic Needs" for Roos (Welfare Theory and Social Policy, Helsinki, 1973, p. 65) means "the imperative for people to be able to live and develop themselves";

(d) Ahmavaara bases his list of basic needs on unconditional reflexes (Ahmavaara, Yhteiskuntatieteen kyberneettinen methodologia ja metodologisen posetivismin kritikki, Helsinki, 1970, pp. 134-136;

(e) for Amitai Etzioni basic human needs are those specific to man - The Active Society, a Theory of Societal and Political Processes, New York, Free Press, 1968, pp. 624-626.

(7) In this connexion it is interesting to note how easily basic ideas and notions in liberal production theories have been assimilated in Marxist theory and vice versa - at least as far as these ideas relate to how production and profit can be increased (instead of "profit" other terms such as "economic way of production" are used in Eastern Europe).

(8) The alternative would be more holistic ways

of dealing with the human condition - possibly leading to forms of understanding so complex that the action consequences may be negligible. The Western form of understanding gives power to the professional competent in the handling of a limited range of variables and not to the contemplative person who by intuition feels his way in very complex webs of units and variables. Marxism takes an in-between position: very simplistic in terms of the basic causal variables, but very complex in its perception of the totality affected by that variable.

(9) In either perspective the inner limits of man and the outer limits of nature are easily neglected or at least not given primacy. More prominent is the rôle played by economic competition (by private or collective entrepreneurs) and by class conflicts (classes of individual and collective actors).

(10) A well-known example of this - and one which has had as great significance in theory as in practice as it was written at about the beginning of the last century, can be found in Adam Smith, An Inquiry into the Nature and Causes of the Wealth of Nations, London, McCulloch, 1839, Ch. 6, p. 22.

(11) This leads to the peculiar idea that capitalism is a stage societies have to go through - in spite of all its anti-human characteristics. See Miklos Molnar, Marx, Engels et la politique internationale, Paris, Gallimard, 1975, pp. 192, 193, 197, 198, 205, 268.

(12) See, for instance, R. Barnet and R. E. Muller, Global Reach, N. Y., Simon & Cluster, 1974. In this excellent study the authors show in detail how the multinational corporations work in order to create needs for their products in developing countries. In order to do this "on a high profitable basis" ethnographers and psychologists participate in the task. Barnet and Muller refer, for instance, to the development patterns in the Latin American countries (Ch. 9, pp. 213-254).

(13) See Johan Galtung, "A Structural Theory of Imperialism", Essays in Peace Research, Vol. IV, 13, Copenhagen, Ejlers, 1977.

(14) This is a basic thesis in the Trends in Western Civilization Programme of the Chair in Conflict and Peace Research, University of Oslo.

(15) Today the ranks of the imitators are rapidly dwindling and not only in the capitalist camp: the meeting of communist parties in Berlin, June 1976, also dethroned the Soviet Union as model country.

(16) The first systematic presentation appeared in 1951, in The Social System, Glencoe, The Free Press. For a discussion, see Johan Galtung, Members of Two Worlds, Oslo 1971, Chapter 1.2.

(17) Substitutability is in line with the general tendency towards increasing compartmentalization of the individual personality. In practice this shows up in the way in which each individual can be seen in a set of rôles, with corresponding rôle expectations, and the system of production is only interested in one of these rôles. This pattern is seen particularly clearly in connexion with imported foreign workers in rich Western countries: one is mainly interested in the "productive part" of these workers, and adjusts conditions to meet the requirements in that connexion. The individual foreign worker develops a feeling of living in a vacuum and a sense of hopelessness when he attempts to use parts of oneself other than those directly related to conditions of production.

(18) Marx is of the opinion that these are among the factors that distinguish human beings from animals - the difference between the least skilful master builder and the best bee, is that the former has built the cell inside his head before it is built in wax - Capital, Ch. 5, first part.

(19) One reason why all these studies were and are still being carried out is no doubt the ease with which it can be done - as in Russell et al. Handbook of Social and Political Indicators, reducing analysis to a mechanical job once the data are there.

(20) For one example of the use of diachronic analysis, see Johan Galtung, "On the Relationship Between Human Resources and Development: Theory, Methods and Data", in N. Baster, ed., Measuring Development: The Rôle and Adequacy of Development Indicators, London, Frank Cass, 1972, pp. 137-153.

(21) For an analysis of Japanese style economic imperialism, see Johan Galtung, "Japan and Future World Politics", Essays in Peace Research, Vol. V, 6, Copenhagen, Ejlers, 1977.

(22) Czechoslovakia could be cited here as a case in point. Formerly part of the Austro-Hungarian Empire, it was neither in the centre nor in the periphery. Of all the East European countries, it was the one for which the Soviet experience was least relevant as well as being the country which, left to develop its own socialism, could also serve as a model for the Soviet Union in some major respects. The events in August 1968 eliminated this possibility. What happened then was probably not entirely unwelcome in the capitalist West because "communism with a human face" also would constitute a threat to these countries with the terror element stressed by Western media removed. Some of the same applies to Italy and France. What the other Western countries fear is not a terror régime, but a basic change in economic

structure combined with respect for human rights and freedom.

(23) This, of course, refers to the People's Republic of China. The question is to what extent it is really inspired by Marxist thinking. For some reflections on this, see Johan Galtung and Fumiko Nishimura, Learning from the Chinese People, Oslo, 1975 (in Norwegian, English edition forthcoming), especially Chapters 2 and 8.

(24) For one discussion of how such criteria may operate in Eastern European societies, see Johan Galtung, "Social Imperialism and Sub-Imperialism: Continuities in the Structural Theory of Imperialism", World Development, No. 1, 1976.

(25) The usual pairings for comparison would be North vs. South Korea; North vs. South Viet Nam up to 30 April 1975 - now, perhaps, Viet Nam or the former Indo-China in general as against the ASEAN countries; China vs. India and Cuba vs. Venezuela or the Dominican Republic.

(26) It cannot be sufficiently emphasized that there is a very basic choice involved here: whether one identifies development with what happens at the top, at the middle, or at the bottom of society. No doubt the focus on averages is a statistical concession to the general population, a step forward relative to the idea of measuring welfare by the standards of a royal court or the upper classes in general. But the point here is not merely the statistically obvious that averages do not reflect dispersions, but the humanly crucial point that it is at the bottom of society that the threshold between well-being and misery is located. Hence the argument should be made that to lift people above this threshold is the key point in development.

(27) One might say that the point of departure for Marx' social criticism and analysis was precisely the idea of putting human beings in the centre. He found a gap between the nature of human beings and their existence. The natural essence of Man was for Marx the ideal which was suppressed due to structural conditions. Empirical Man, as formed by the capitalist system represents only human existence, prevented from realizing his true human nature. See, for instance, Economic-Philosophical Manuscripts, p. 24 in Marx' pagination, Paris 1844, in Marx/Engels Historisch-kritische Gesamtausgabe, Frankfurt, Berlin, Moskva 1927-1935.

(28) One could also imagine a tripartition of these different ways of viewing human beings: ideas about human beings as they really are - based on notions of human nature; ideas about human beings as they appear in a given situation in a given society, the empirical human being; and ideas about human beings as they ought to be,

the imperative image of Man. To illustrate these three aspects let us make use of the concept of alienation which in traditional Marxism usually is based on a positive view of Man, the true human nature, and Man as Man ought to be. Empirical Man, however, is alienated: Freud had a different view: if human beings could really develop as human beings, it would be impossible to build the good society with a corresponding high level of culture; Freud's view is more negative, pessiministic, as opposed to the view held by W. Reich: human beings not only can but must satisfy all their needs, and this is a necessary condition for the good society (Reich, The Sexual Revolution, London, 1969). Needless to say, ideologies and viewpoints with a negative view of man can easily be used in order to promote an authoritarian social system in order to domesticate, control and repress what is held to be the true but, at the same time, a threatening human nature.

(29) Different views of Man are expressed in the various suggestions as to how to characterize Man as a genus: homo faber, homo sapiens, homo ludens, homo negans, homo desperans - as explored by Fromm in The Revolutions of Hope, New York, 1968, World Perspectives No. 38, Part 4 - What does it mean to be human?

(30) Thinking of the psychologists who some years ago were exploring "human instincts" and finally thought they could locate no less than 6,000 of them with each theoretician adding his own suggestions, many would be in favour of very short lists of needs, focusing on the most basic ones. We are, however, of the opinion that it might also be interesting to focus attention on suggested instincts about which there was a certain consensus, in other words, on the degree of overlap in answers. In addition, there is the advantage to a relatively long list of needs that it runs a better chance of covering more of the total personality. About instincts and the development towards several thousands of them, see, for instance, E. Murray, Motivation and Emotion, New York, Englewood Cliffs, 1964.

(31) The idea that human needs vary over time should be qualified by making a clear distinction between quantitative changes (people need more of the same to be satisfied) and qualitative changes (people need new types of satisfaction, not only new levels). Incidentally, another aspect of time would be how the needs vary with age. The infant, the child, the adolescent, the adult and the old may all be said to emphasize different types of needs. One interesting thesis in this connexion would be: if one could make a list of the relatively specific needs characteristic of the age interval from 20 to 60 years, then one could also

make a survey of the processes leading to the satisfaction of precisely these needs. One hypothesis would then be that in the over-industrialized societies precisely these processes are given high priority because they lead to the satisfaction of needs for the age group in power - particularly the male part of the middle aged.

(32) In the literature on needs different criteria are emphasized, some of these are mentioned in note 6.

(33) We would stress "at least to some extent" as it is unacceptable to insist on a specific hierarchical arrangement of needs in all cases as Maslow could be said to do - in Towards a Psychology of Being, New York, D. van Nostrand, 1968.

(34) The only set of needs about which there seems to be a consensus that they are more basic than others are those that constitute necessary conditions, physiologically speaking, for remaining alive.

(35) We find the basis for this assertion in the current world situation, characterized by lack of satisfaction of the most basic material needs in so many parts of the world. Do the various declarations on human rights mirror this particular aspect of the world situation? We think they do not, or at least only insufficiently even though one might say that the United Nations Declaration of Human Rights has certain paragraphs that are related to basic material needs:

para. 3:   Everyone has the right to life, liberty and security of person.

para. 23.3: Everyone who works has the right to just and favourable remuneration ensuring for himself and his family an existence worthy of human dignity and supplemented, if necessary, by other means of social protection.

In addition, one might also mention paras. 24, 25 and 26.1. What is missing, however, is explicitness with regard to the material aspects of these needs: "human dignity" being a different kind of concept than food, clothes, shelter, health and education.

(36) It is sufficient to recall the French declaration of 1789. But then again it seems easier to point to the weaknesses in the declarations of the past than in the declaration of the present, hardly because the latter necessarily are better, but because we do not see their weaknesses so clearly.

As George Lefebvre points out in his Quatre-vingt-neuf, Paris, 1939, it was a declaration stipulating a set of norms for the most powerful society - in sharp contrast with the reality of French society at that time, with hundreds of thousands living on, or below,

the starvation line, certainly not satisfying their most vital and basic needs and rights. "Needs and rights": the general idea would be that needs should become institutionalized as rights, and only those rights that correspond to needs should be recognized.

(37) Indeed in times of war, human rights are violated almost every day in the world. As to accidents: it should be unnecessary to emphasize that in developed and over-developed societies the right of life is not at all satisfactorily protected - particularly in the field of transportation where in many countries the children are over-represented as victims.

(38) Clearly, this criterion - that people have been fighting for these rights - cannot cover everything. The most suppressed will not have the necessary material, spiritual and human resources in general to fight for what others might conceive of as an evident human right - they may, for instance, be too short on information.

(39) Maslow's hierarchical needs are as follows: First, the somatic needs, such as:
1. Physiological needs
2. Security needs, and then
3. Needs for solidarity, context and acceptance
4. Need for self-respect and status
5. Need for self-realization
6. Need to know, learn, discover
7. Need to symmetry, beauty, aesthetical qualities

(here quoted from Roos, op. cit., p.68). It should be pointed out that the "need for symmetry" may be culturally very specific: as in contrasting a Western and a Japanese garden, the former is often symmetric (Versailles), while the latter is often organized around a central point, which may not be located inside the garden.

(40) Many might add that it is exactly in the unfinished, where there remain many important things to do. Individually or collectively, a necessary condition can be found for human well-being. In his preface to the second edition of Motivation and Personality Maslow emphasizes precisely this need for the unfinished and the imperfect. He says, for instance: "The demand for 'Nirvana Now!' is itself a major source for evil, I am finding" (p. XXII - Harper & Row, 1970).

(41) Can capitalism be non-expansionist? Can it just lock in at a point of stable equilibrium and stay there? Or, will it always be driven onwards through the combined ideology provided by comparative advantages and economies of scale, into ever-expanding systems? One important consideration here would be that this does not only depend on "laws" of economics, in casu of capitalist economics, but also on cultural, even civilizational variables.

Thus, in a less expansionist civilization than the West, capitalist social and economic formations might perhaps stay within certain borders, e.g., within the confines of a local or a national economy, and in stable equilibrium. Stability may become institutionalized just as expansion may; it is difficult to see how one or the other is the necessary consequence of a capitalist economy. It may even be that those who have so argued have committed a major mistake in disregarding civilizational variables.

(42) See footnote 13.

(43) See footnote 24.

(44) In the general theory of needs one might refer to the distinction between the need subject - the individual human being - and the need object, non-material or material conditions that can be utilized for the satisfaction of needs. A concrete tool can from this point of view be regarded in at least two different ways: as a tool to produce need objects, for instance food, or the tool can in and by itself be a need object, e.g., because a need to be creative, active, to work is satisfied when the tool is made use of. For the distinction between need subject and need object, see J. P. Roos (op. cit., p. 65) or Anders Wirak, "Human Needs as a Basis for Indicator Formation", Papers, Chair in Conflict and Peace Research, University of Oslo, 1976.

(45) These problems have recently been analysed in many good studies, and also in exhibitions, such as the "Ararat" exhibition in Stockholm, summer 1976 (at the Modern Museum) where the theory has been put into practice in the sense that concrete products (need objects) more in harmony with nature have been created - from forms of human habitat, to human and nature friendly energy and production systems.

Spatial Patterns and Regional Structures in Thailand: an Application of Territorial Indicators as an Input in the Development Planning Process*

R.G. Cant

(Department of Geography, University of Canterbury, Christchurch, New Zealand)

* Unesco workshops document No. SHC-75/WS/57, dated November 1975.

SPATIAL PATTERNS AND REGIONAL STRUCTURES IN THAILAND:
AN APPLICATION OF TERRITORIAL INDICATORS AS AN INPUT IN THE
DEVELOPMENT PLANNING PROCESS

R. G. Cant

This study is a work-in-progress document presented at the point where two research programmes converge. National planners in Thailand have widened their terms of reference and are expanding their evaluation and planning machinery to embrace social and spatial concerns. At the same time the Unesco Programme of Research on Social Indicators has been given a series of new thrusts, one of which is intended to relate the methodology of social indicators to the practical tasks of national planning. The material set out here is intended as a basis for dialogue between planners and policy-makers on the one hand, and research workers and academics on the other.

SOCIAL AND ECONOMIC DEVELOPMENT PLANNING IN THAILAND

Development planning in Thailand has made rapid advances in the course of a brief history. The first formal initiatives were taken in 1959 following discussions between the Thai Government and international lending institutions. The National Economic Development Board was set up and given the task of formulating the First National Economic Development Plan to cover the five-year period from 1961 to 1966. Both this first plan, and the Second Economic Development Plan (1967-1971), placed particular emphasis on such sector activities as transport and education and various forms of production infrastructure such as irrigation schemes or electric power. Substantial advances were made in almost all of the target sectors.

On the basis of these initial achievements, and in the light of new needs and a wider social awareness, the third and fourth plans have been enlarged in their terms of reference. There is a new emphasis on social conditions and a firmly stated desire to improve the quality of life of the Thai people both in the cities and in the rural areas. The development policies for the Third Plan (1972-1976) indicate how closely social and economic considerations are interwoven

in the eyes of the planners and policy-makers.

(1) "To restructure the economic system and to promote economic growth.

(2) To maintain economic stability.

(3) To promote economic growth in the rural areas and to reduce income disparities.

(4) To promote social justice.

(5) To develop manpower resources and to create employment.

(6) To foster the rôle of the private sector in economic development." (NESDB, 1971, vi)

The major objectives of the plan, as set out by the Prime Minister in his promulgation address, were to "develop human resources simultaneously with natural resources", partly because "human resources play a leading rôle in the effort to increase national productive capacity", and partly because "the development of human resources is a means to promote and maintain the desired peaceful and stable society". Special attention was given to the human resources of rural areas since "the efficiency of the rural labour force is very closely related to raising incomes and living standards of rural people, which will eventually contribute to achievement of social justice" (NESDB, 1971 vi). This widening of concern is acknowledged by the fact that the National Economic Development Board has been renamed the National Economic and Social Development Board and the plans are now known as Economic and Social Development Plans.

A second new dimension for planning, which commenced in the Second Plan and received new impetus in the Third Plan, is an explicit intention to spread development efforts so that they reach the rural population in the various regions and provinces. Regional planning of this kind was a new venture for Thailand and it was initiated in stages, beginning with the strategic peripheral regions of the north-east and north and then extending down the administrative hierarchy to the 15 changwats (provinces) of the north-east region (NESDB, 1971, 87). At the time of implementation of the Third

36

Plan (1972-1976) it was intended that regional planning should be extended to south and west Thailand as well. South Thailand was recognized as a region with "a distinct geography, and a distinct economic and social system, particularly along the border provinces" and the main policy here as elsewhere would be "to accelerate economic and social development". (NESDB, 1971, 90)

At the present time, then, we have a situation where the social and spatial dimensions of planning are clearly recognized and the planning staff of the National Economic and Social Development Board are searching for the technical means whereby they can evaluate regional needs and opportunities, identify appropriate strategies, and measure the progress which is being made towards the achievement of planning goals.

THE SOCIAL INDICATORS PROGRAMME OF UNESCO

The Division of Social Science Methods and Analysis* of Unesco has worked on systems of human resource indicators from 1967 onwards. In a number of theoretical and methodological studies they have used such indicators to measure levels of development and explore the interrelationships, between human resources and development (Unesco, 1975a, 195-6).

In 1974, two new projects were initiated; one concerned with the identification of key indicators of social and economic change, the other intended to relate the use of socio-economic indicators to the practical task of development planning and, in particular, to the problems involved in the identification and elimination of social and economic inequalities. In keeping with its intention of relating the theoretical and methodological advances to practical planning a series of regional working groups were organized. One such meeting, held in Bangkok in September 1974 drew together participants from seven Asian countries, from Unesco itself, from the United Nations Institute for Economic Development and Planning in Bangkok, and the Thai National Economic and Social Development Board.

The Bangkok workshop was faced with two practical questions:

"(1) Is it feasible to use socio-economic indicators for national planning?

(2) What type of indicators are appropriate?" (Unesco, 1975b, 1)

The papers and discussions at the workshop focused on the use of indcators at both the national and sub-national or regional levels. As a contribution to the latter aspect of the discussions, the present author examined the rôle of territorial socio-economic indicators in the formulation and implementation on national plans (Cant, 1974). The methodology discussed in that paper is now taken and applied for Thailand.

TASK AND METHODOLOGY

If the planner is to devise and implement development strategies which will meet the varied needs and develop to best advantage the human and physical resources of each region and province, he must be able to identify, summarize and display the pertinent characteristics of each territorial unit. He needs to be able to measure mixtures of economic activity and levels of social welfare in all parts of Thailand and, as far as possible, to be able to relate these to the natural resources and the existing infrastructure of each area. In other words, he needs to map the main features of the social and economic landscape and identify the development strengths and weaknesses of particular regions and changwats. To carry out this task he needs three things: firstly, a framework for spatial analysis, secondly, a series of economic, social, demographic and environmental indicators and, thirdly, techniques of analysis that will enable him to identify significant relationships and map important spatial variations.

The administrative framework which has evolved in Thailand is a particularly convenient one for spatial analysis. There are 71 changwats, all of similar order of size in either area or population, and each containing a municipal administrative centre and a surrounding rural hinterland. Only two, Phra Nakhon and Thon Buri, are mainly urban and none is completely rural. Census data embracing a wide range of economic, social and demographic items are published separately for each changwat and additional environmental and economic data for the same units have been tabulated in the National Resources Atlas.

The Atlas and the Census both recognize a number of separate regions: those recognized by the National Resources Atlas are shown in Figure 1. In the present analysis the 71 changwats form the basic building blocks, but where appropriate the results can be mapped or graphed for smaller portions of the country.

In selecting indicators for an analysis of this type, an attempt should be made to strike a balance between the various areas of concern that are relevant to the planning process and to avoid, as much as possible, indicators which by their very definition are known to correlate (e.g., incomes of heads of household and income of households). The indicators chosen for this study are discussed in more detail below. At this stage it is sufficient to note that adequate data were available from the 1970 Population and Housing Census to select a range of economic, social and demographic indicators. In addition it was possible to add a limited number of environmental and land use indicators from the information tabulated in the National Resources Atlas.

---

* Restructured on 1 April 1976 as the Division for Socio-economic Analysis.

Fig. 1 - THAILAND: CHANGWATS AND NATIONAL RESOURCE ATLAS REGIONS

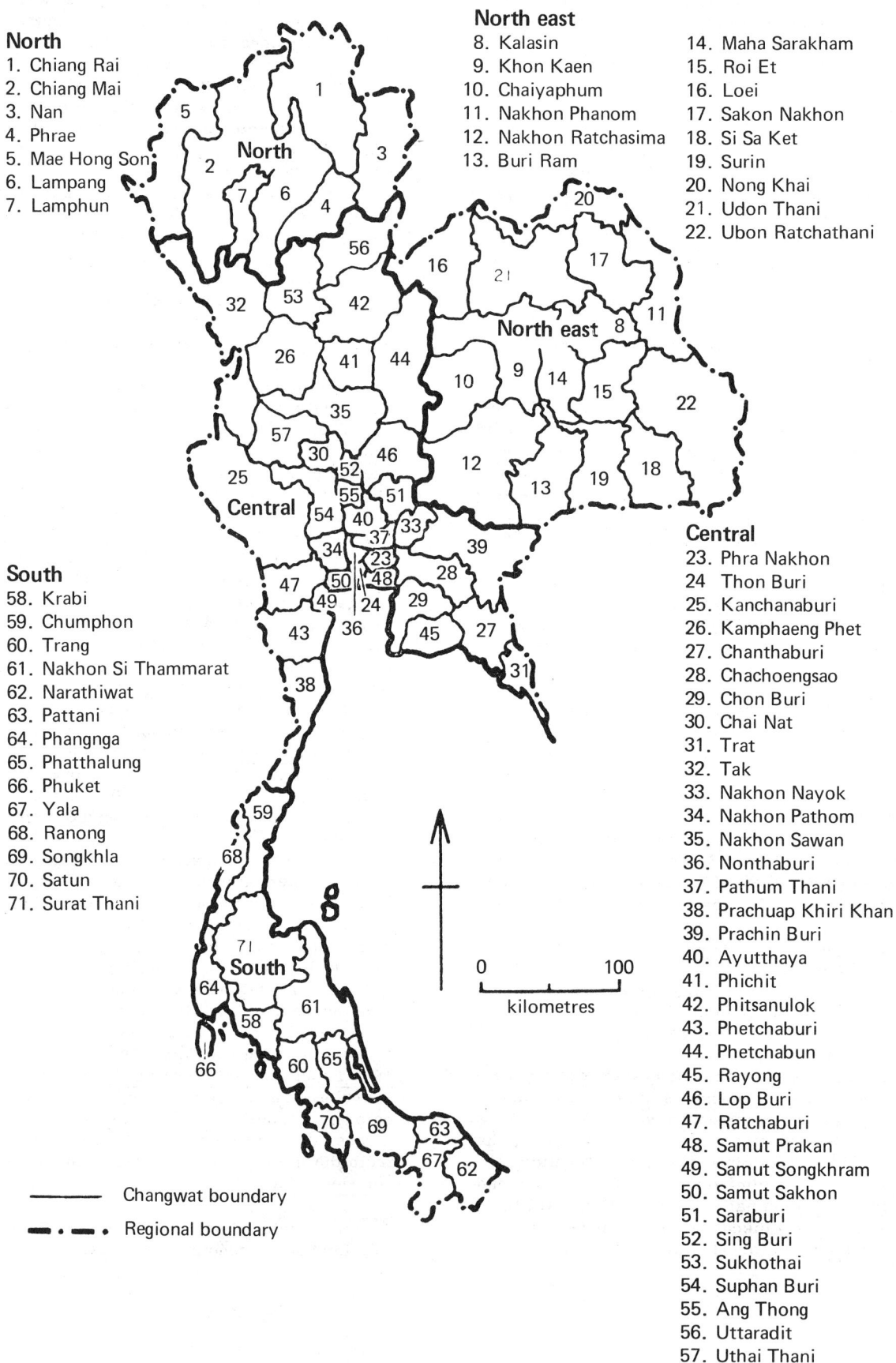

North
1. Chiang Rai
2. Chiang Mai
3. Nan
4. Phrae
5. Mae Hong Son
6. Lampang
7. Lamphun

North east
8. Kalasin
9. Khon Kaen
10. Chaiyaphum
11. Nakhon Phanom
12. Nakhon Ratchasima
13. Buri Ram

14. Maha Sarakham
15. Roi Et
16. Loei
17. Sakon Nakhon
18. Si Sa Ket
19. Surin
20. Nong Khai
21. Udon Thani
22. Ubon Ratchathani

South
58. Krabi
59. Chumphon
60. Trang
61. Nakhon Si Thammarat
62. Narathiwat
63. Pattani
64. Phangnga
65. Phatthalung
66. Phuket
67. Yala
68. Ranong
69. Songkhla
70. Satun
71. Surat Thani

Central
23. Phra Nakhon
24. Thon Buri
25. Kanchanaburi
26. Kamphaeng Phet
27. Chanthaburi
28. Chachoengsao
29. Chon Buri
30. Chai Nat
31. Trat
32. Tak
33. Nakhon Nayok
34. Nakhon Pathom
35. Nakhon Sawan
36. Nonthaburi
37. Pathum Thani
38. Prachuap Khiri Khan
39. Prachin Buri
40. Ayutthaya
41. Phichit
42. Phitsanulok
43. Phetchaburi
44. Phetchabun
45. Rayong
46. Lop Buri
47. Ratchaburi
48. Samut Prakan
49. Samut Songkhram
50. Samut Sakhon
51. Saraburi
52. Sing Buri
53. Sukhothai
54. Suphan Buri
55. Ang Thong
56. Uttaradit
57. Uthai Thani

North
North east
Central
South

0   100
kilometres

———— Changwat boundary

—·—·— Regional boundary

The methodology used is discussed in some detail in a previous publication (Cant, 1974). In brief, the procedure followed was to prepare a data matrix containing values for 71 changwats and 32 selected indicators. The technique of principal components analysis was used to identify clusters of relationship among the 32 variables and present the most important of these in the form of orthogonal factors (as in Table 1) which are given verbal interpretation. As an extension of the principal components analysis, the original indicator values for each changwat are used to compute new factor scores of each factor on the same changwat. These new scores are mapped or graphed and the patterns interpreted accordingly.

This visual display of the new factor scores is presented selectively in the form of a nested set of generalizations. At the first level a series of maps for Thailand as a whole gives a broad over-view of the national space economy (Figure 2). At the second level the focus shifts to two of the regions and the scores for two of the factors are plotted in a scattergram which relates levels of economic development and well-being in these particular regions to those enjoyed by Thailand as a whole (Figure 3). At the lowest level of resolution, factor scores are used in the preparation of changwat profiles which enable us to highlight the strengths and weaknesses in levels of development for each changwat in one particular region (Figure 4).

Where appropriate in the analysis an alternative methodology is applied. Instead of the correlative technique which is basic to the principal components analysis an additive technique is introduced. In this case clusters of indicators are selected and given equal weight by simply standardizing the values for each indicator and adding the standardized scores to form new composite indices. Once the addition is completed these composite indices are standardized to a mean of zero and a standard deviation of unity. They can be interpreted with less ambiguity and are thus used to verify or extend conclusions used on the basis of principal components analysis (Figure 5).

THE THAI DATA MATRIX

The indicators used in the study do not cover the full range of published statistics but are drawn from two major sources which provide changwat level information for 1970. The first of these is the 1970 Population and Housing Census (changwat series in 71 separate volumes) and the second is the National Resources Atlas published by the Royal Thai Survey Department in 1972, but containing maps and tabulations compiled by different agencies at various dates prior to date of publication (see Appendix 1). In other words, the analysis presented here provides a base-line study on conditions in Thailand at the beginning of the present decade.

The indicators were selected and the data needed to compute values for each index in each changwat were recorded in a 71 by 32 data matrix. The aim has been to use a variety of indicators which span across fields of activity which are considered pertinent to the planning process and reflect some of the spatial goals set out in the Third Development Plan. Such social indicators as housing, education and (un)employment receive special emphasis, while others in areas such as health and incomes would have been included had the necessary data been available. The indicators selected group into four broad classes:

(a) Environmental resources and land use indicators;

(b) Economic indicators (employment by industries);

(c) Social welfare indicators (education, employment, housing and amenities);

(d) Demographic indicators.

In the case of indicators for housing and amenities it was possible to obtain separate data for municipal and village populations and thus calculate these items separately for each changwat. An attempt to calculate the number of medical workers per 10,000 population was abandoned when it was found that the Census did not publish a full occupational breakdown for all changwats. A full listing of the indicators used, together with definitions and sources, is given in Appendix 1.

In specifying and selecting the particular indicators used in the analysis some subjective decisions have been made which will influence the form of the results. These should be clearly stated at this point and kept in mind as we proceed with the interpretation. To begin with, the fact that there are only five environment and land use indicators as compared with 15 social welfare indicators will tend to produce results which will give greater emphasis to levels of living than to environmental or land use variations. In contrast to this first situation we have only four economic indicators but in this instance the four items concerned - male employment in agriculture, forestry and fishing; male employment in mining and quarrying; male employment in manufacturing and male employment in transport and commerce - are all divided by total male employment. We have, by the choice of this denominator, closed the number system for these variables and introduced a cluster of intercorrelations which will give an explicit economic structure to at least one factor in the first stage of the analysis.

SPATIAL VARIATIONS IN ECONOMY
AND WELFARE

In this first stage the aim is to isolate important sources of variation in economy and welfare and map the broad patterns of the Thai space economy.

The computer programme used, a principal components analysis with varimax rotation, calculates factor loadings for a selected number of factors and, in addition, calculates the standardized scores for each changwat on these new factors. In this instance the results with three, four or five factors were all computed and examined before a decision was made to focus the discussion on the four-factor solution shown in Table 1. The factor scores, which are available in Appendix 2, have been used to produce the maps shown in Figure 2.

Each of these factors can be interpreted by means of the factor loadings shown in the columns of Table 1 which can be regarded as measures of the correlation between particular factors and particular variables. In factor 1, for example, there are strong positive loadings with employment in manufacturing (variable 8, loading 0.86) and employment in commerce and transport (variable 9, loading 0.93). The same factor has a strong negative loading for variable 6 which is employment in agriculture, forestry and fishing. In other words, this first factor distinguishes between territorial units where primary employment is of greater importance and areas where secondary and tertiary employment are more important. The remaining loadings in column 1 additional information about these economic differences and the social and demographic differences that are associated with them.

Factor 1 can thus be designated as an employment factor which distinguishes between more rural areas where primary employment is more important and more urbanized areas where secondary and tertiary activities are dominant. In the Thai situation it is evident that the latter areas with a more diversified employment structure tend to provide better educational facilities at secondary level and at the same time record much higher levels of unemployment among new workers. Rural agricultural areas tend to have higher levels of general fertility and a higher proportion of child dependents. Rural houses in more urbanized changwats are more likely to have piped water and the households which live in them are more likely to possess radios.

When we map factor 1, as in Figure 2(a), we identify a very clear core and periphery contrast. The commercial - industrial core focuses very firmly on Bangkok and the coastal areas of the Central Plain while the primary producing periphery incorporates the whole of the north east region together with a number of provinces in the interior of the Central Plain and in the north region. The map discussed here sketches in bold, simple terms a pattern of relationships that is well-known to planners and research workers alike.

A second source of variation, uncorrelated to that just discussed, is identified in factor 2 which has most of its higher loadings on variables 17 to 20, all measuring urban household amenities. Changwats in Thailand thus appear to differ in the standard of urban amenities quite independently

of their employment structure. The loadings also imply that changwats which are better off in terms of urban amenities tend to have higher levels of literacy and attract migrants from other changwats. Factor 2, which we will call an urban amenities factor, thus identifies one component of social welfare and enables us to map areas which are advantaged and disadvantaged in this particular respect.

When we map factor 2, as in Figure 2(b), we find that areas with urban amenities above or below the normal for Thailand as a whole are widely dispersed. Changwats which include greater Bangkok, for example, score particularly well while some other changwats immediately adjacent including Pathum Thani, Chachoengsao, Ayutthaya, Samut Sakkon and Samut Songkhram record low scores. There is considerable diversity in the north-east where a number of changwats score more highly on this factor than on the previous factor. At the southern extremity of the country adjacent to Malaysia, eight out of twelve provinces record very low scores.

A third basic contrast, this time a land use contrast, is identified in factor 3 which separates out areas where riceland is more dominant from areas where forest land is more important. Associated with these land use differences are certain differences in material welfare; rice growing areas tend to have more of their urban houses built of strong materials and a higher proportion of their village households report radios and bicycles. Forest areas in Thailand show similar frontier characteristics to their counterparts in other countries; they have a higher proportion of males in the population and fewer aged dependents. The fact that unemployment among experienced workers tends to be higher in riceland areas (variable 14) may either reflect the older age structure of the population in these areas or it may result from the seasonal nature of rice cultivation and the timing of the 1970 Census.

When factor 3 is mapped in Figure 2(c), the Central Plain stands out as the predominant area for rice cultivation. Forest is most dominant in the northern and western extremities of Thailand and appears to be important in some parts of the north east. There is, however, an element of ambiguity about the factor scores. The values recorded for the two metropolitan changwats of Phra Nakhon and Thon Buri derive from very small values for riceland per capita, rather than the presence of forest. They must thus be regarded as an artefact of the computation process rather than an accurate indicator of the character of the changwats concerned.

The fourth factor has positive loadings on tree crop land and on mining and quarrying. It is less important in national terms, but when the positive scores are mapped it highlights the special character of south Thailand and the southeast portion of the central region (Figure 2(d)).

Table 1: Thailand: Spatial dimensions of economy and welfare, 1970[1]

Variable	Factors[2] 1	2	3	4
Environment and land use				
(1) Forest land/capita			-0.60	
(2) Farmland/total land	0.63		0.53	
(3) Riceland/capita			0.70	
(4) Tree crop land/capita				0.71
(5) Upland crop land/capita				
Employment by industries				
(6) Agriculture, forestry, fishing	-0.92			
(7) Mining and quarrying				0.69
(8) Manufacturing	0.86			
(9) Commerce and transport	0.93			
Education and employment				
(10) Primary attendance index				
(11) Secondary attendance index	0.91			
(12) Females at secondary school	0.60			
(13) Literacy		0.51		
(14) Unemployment-experienced		-0.30	0.48	0.40
(15) Unemployment-new workers	0.88			
Household amenities – urban				
(16) Houses of strong materials		0.40	0.66	
(17) Households with electricity	0.37	0.67		
(18) Households with piped water		0.59	0.43	
(19) Households with a car		0.77		
(20) Households with refrigerator		0.73		
Household amenities – rural				
(21) Houses of strong materials				
(22) Households with piped water	0.69	0.50		
(23) Households with bicycle			0.54	
(24) Households with a radio	0.45		0.70	
Demographic				
(25) Municipal population/total population	0.87			
(26) Population density	0.83			-0.31
(27) Male-female ratio		0.40	-0.48	0.66
(28) General fertility	-0.50			
(29) Child dependency	-0.73			
(30) Aged dependency	0.31	-0.61	0.42	
(31) Migrants within changwat				0.48
(32) Migrants from outside changwat	0.37	0.60		
Variance accounted for (%)	26.82	13.18	11.49	8.56

(1) Principal components analysis with varimax rotation of four factors. Loadings below 0.30 are not tabulated. Loadings above 0.50 are underlined.

(2) Factor 1 = Employment structure (secondary/tertiary versus primary); Factor 2 = Urban amenities; Factor 3 = Land use (riceland versus forest); Factor 4 = Tree crops and/or mining.

(3) Calculated from male employment in industrial group/total male employment.

Fig. 2 - THAILAND: SPATIAL PATTERNS OF ECONOMY AND WELFARE, 1970

2a Employment structure

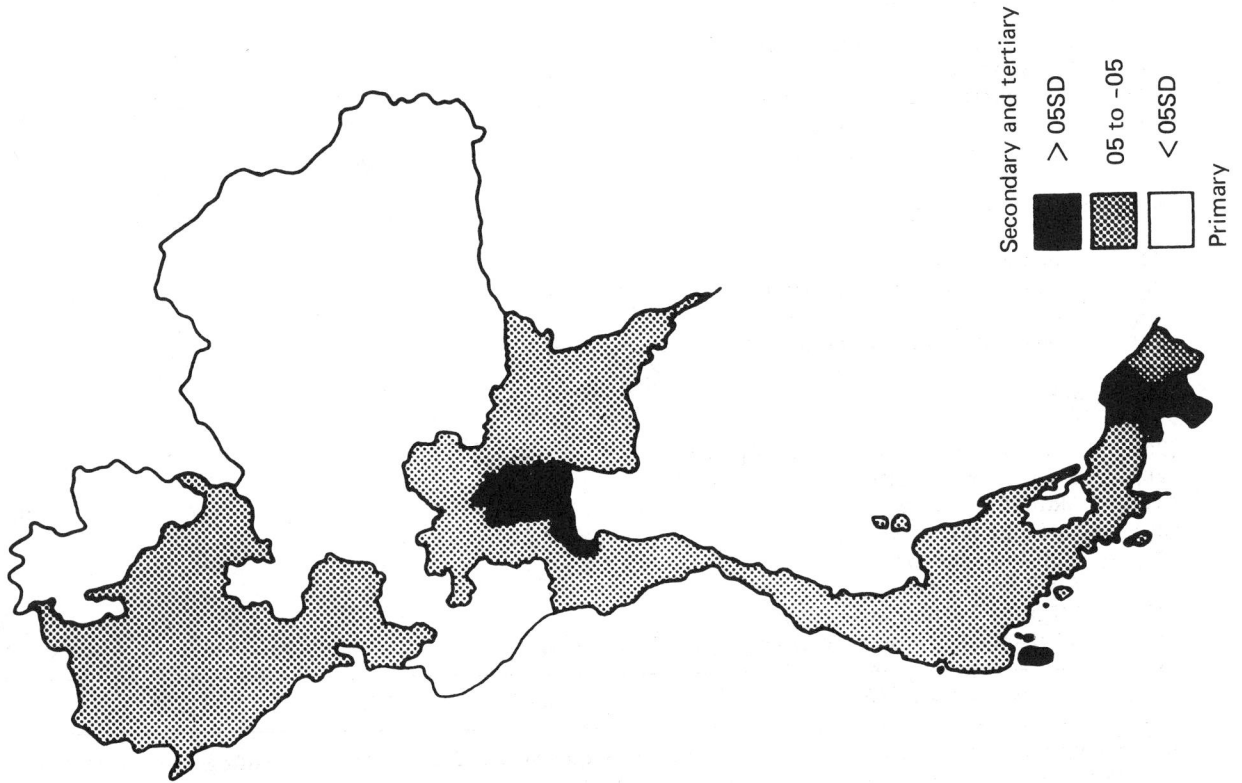

2b Urban amenities

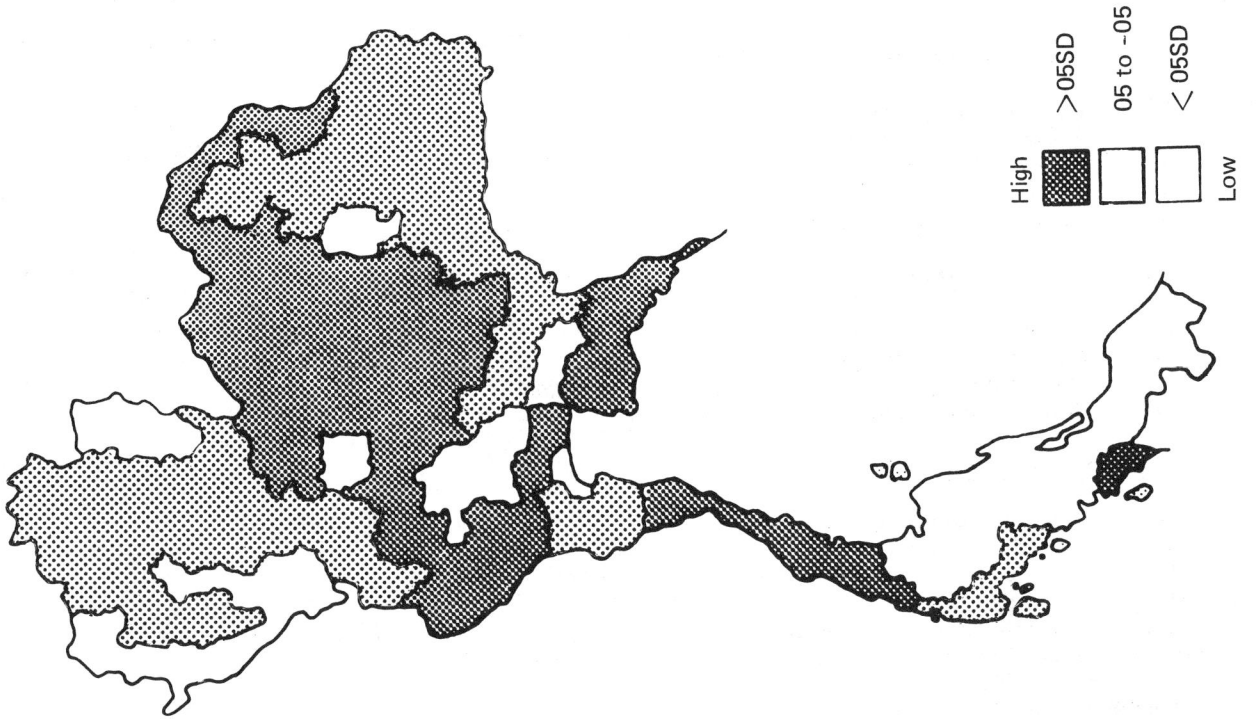

Secondary and tertiary	
■	> 05SD
▦	05 to -05
□	< 05SD
Primary	

High	
▦	> 05SD
□	05 to -05
□	< 05SD
Low	

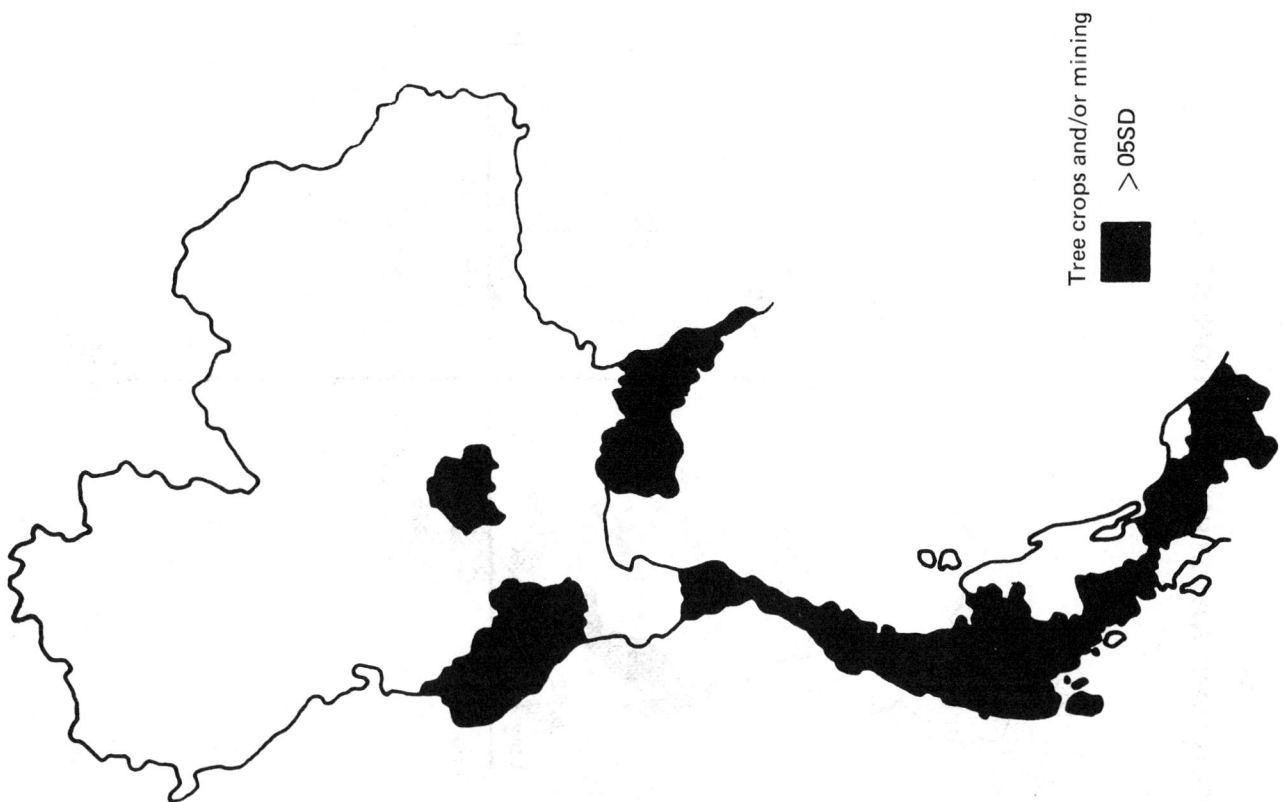

2d Tree crops and/or mining

Tree crops and/or mining

> 05SD ■

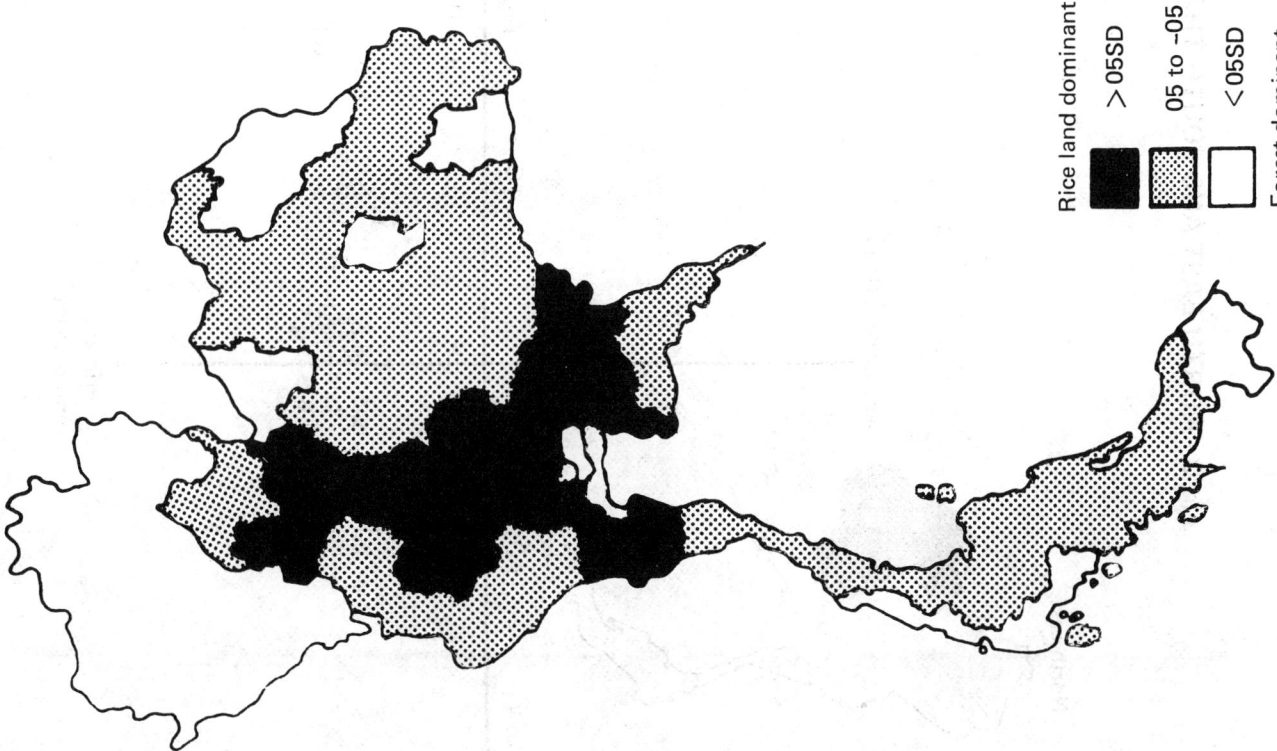

2c Land use

Rice land dominant

> 05SD ■

05 to –05

< 05SD

Forest dominant

43

Fig. 3 - NORTH EAST AND SOUTH THAILAND: CHANGWAT STRUCTURE COMPARED

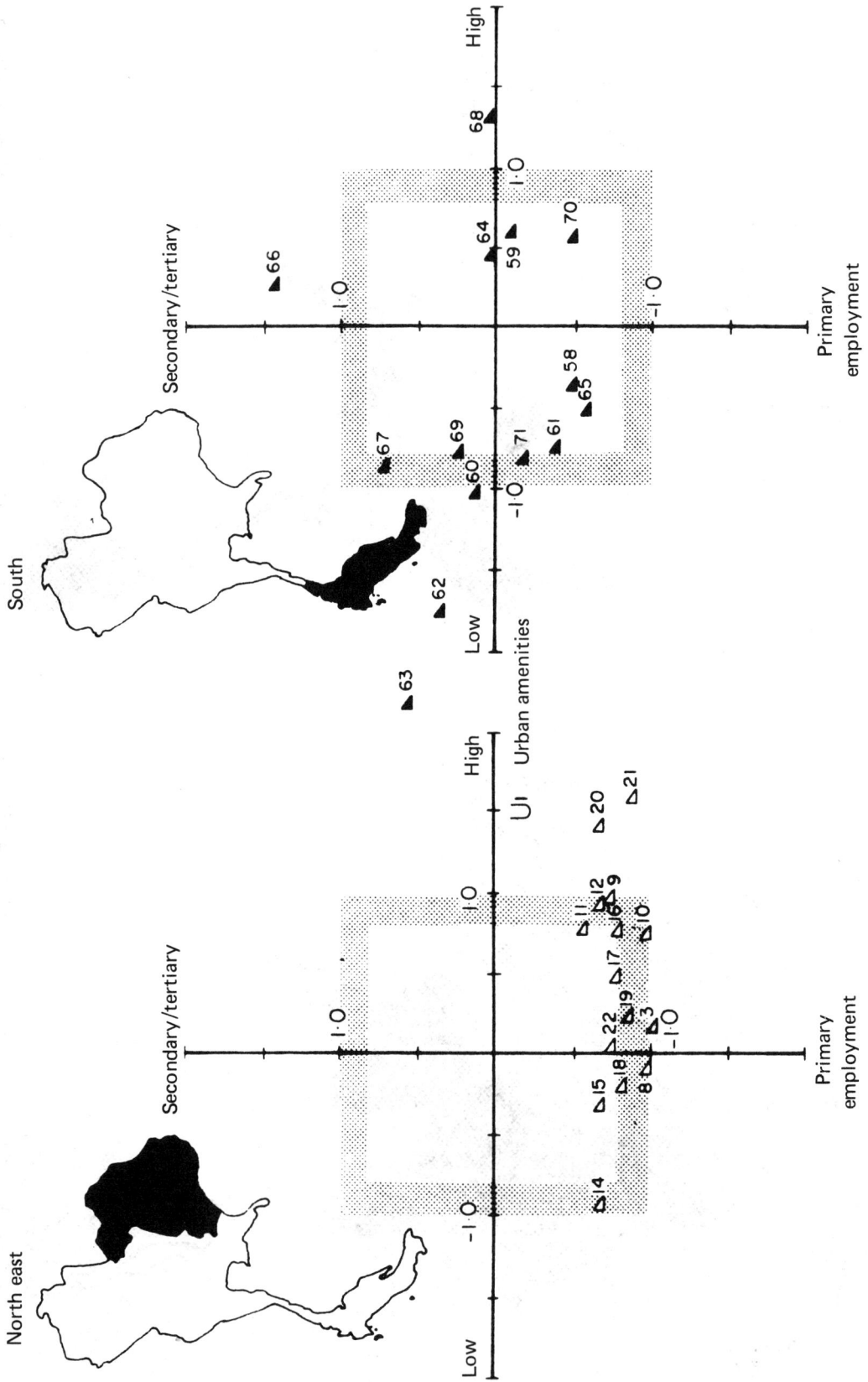

Tin mining and rubber cultivation are important activities in south Thailand and the cultivation of rubber and other tree crops is important in the area to the south-east of Bangkok. In demographic terms, these areas are characterized by a tendency towards male dominance and a greater amount of local migration (variables 27 and 31).

In summary then, the results presented in Table 1 and Figure 2 highlight three major sources of variation in economy, environment and welfare and identify a fourth source of variation which gives special character to one or two peripheral regions. Both in terms of employment and land use there are strong core periphery contrasts but the patterns of these are not identical. In the case of urban amenities, there are broad national variations but considerable local level diversity within individual regions. We now use the same numerical output to focus attention on the particular character of two contrasting regions.

SPECIAL CHARACTER OF NORTH-EAST THAILAND AND SOUTH THAILAND

One of the special concerns of development planning in Thailand, especially in the Third and Fourth Plans, is to stimulate economic growth in peripheral areas and reduce regional disparities in income and welfare. Regions recognized to have special problems include the north-east, made up of the Khorat Plateau and lowlands adjacent to the Mekong River and the Laos border, and the south, made up of a long narrow peninsula which widens out towards the Malaysian border. In the second part of the analysis, selections of output from the previous stage are presented in the form of scattergrams which enable us to contrast conditions in these regions and compare them with what was normal for Thailand as a whole at the time when the data were recorded.

The two most pertinent dimensions available from the results discussed above are employment and urban amenities (factors 1 and 2). In a later stage of the study we will devise an alternative measure which will look separately at urban and rural amenities. In Figure 3 the factor scores for each changwat in these two regions are plotted on axes which measure employment and urban amenities and scale these with reference to the mean and standard deviation for Thailand as a whole. Employment structure is plotted on the vertical axis for each region and the standard of urban amenities is plotted along the horizontal axes so that changwats with better than average urban amenities appear to the right of each scattergram. If all 71 changwats were plotted on the same axes, they would scatter evenly around the origin and most would fall within the central box demarcated by one standard deviation from the mean.

The changwats of the north-east region are all very similar in terms of their employment structure with very little development of secondary and tertiary employment evident in 1970. In terms of urban amenities there is a much wider variation from Maha Sarakham (changwat 14) where amenities are well below the norm for Thailand, to Nong Khai and Udon Thani (changwats 20 and 21 both close to the Laotian capital of Vietiane) where the level or urban amenities is superior to that of most other provinces.

In south Thailand the patterns are much more complex with a wide scatter of scores which cross both axes. Some changwats such as Pattani (63) and Narathiwat (62), both in the extreme south adjacent to Kelantan in Malaysia, have extremely poor urban amenities in spite of a more balanced employment structure while only one, Phuket (66), has a good level of urban amenities and an employment structure with a strong secondary and tertiary employment component. The best level of urban amenities appear to be those of Ranong (68) which is very close to the average for Thailand in its employment structure. It is clear from this comparison that development planning for south Thailand is a much more complex matter than it is for the north-east. In the case of south Thailand differences between changwats often outweigh the broad regional similarities that characterize the north east. For this reason we now turn to south Thailand and examine the individual changwat profiles in greater detail.

THE CHANGWATS OF SOUTH THAILAND, 1970

In Figure 4, profiles have been prepared for each changwat by graphing its scores for each of the four factors identified and discussed in the first stage of the analysis. The scale units for each graph are standard deviations from the Thai mean. Surat Thani (71), the first of the two examples enlarged in the key, is thus distinctive on two counts - its urban amenities are poor by Thai standards while mining and tree crop cultivation are important. The second example, Pattani (63), has even poorer urban amenities but in contrast to Surat Thani, its employment structure is weighted towards secondary and tertiary occupations.

As indicated earlier in the discussion, mining and tree crops are characteristic economic activities in most southern changwats. The graphs show that this is particularly so in the five west coast changwats from Ranong (68) to Trang (60) and also in Yala (67). Only one province, Phatthalung (65) records a negative score.

The scores for the other three factors and the resulting profiles vary markedly from changwat to changwat. In some cases such as Krabi (58) poor urban amenities are associated with forest or frontier conditions and a lack of secondary or tertiary employment opportunities. In

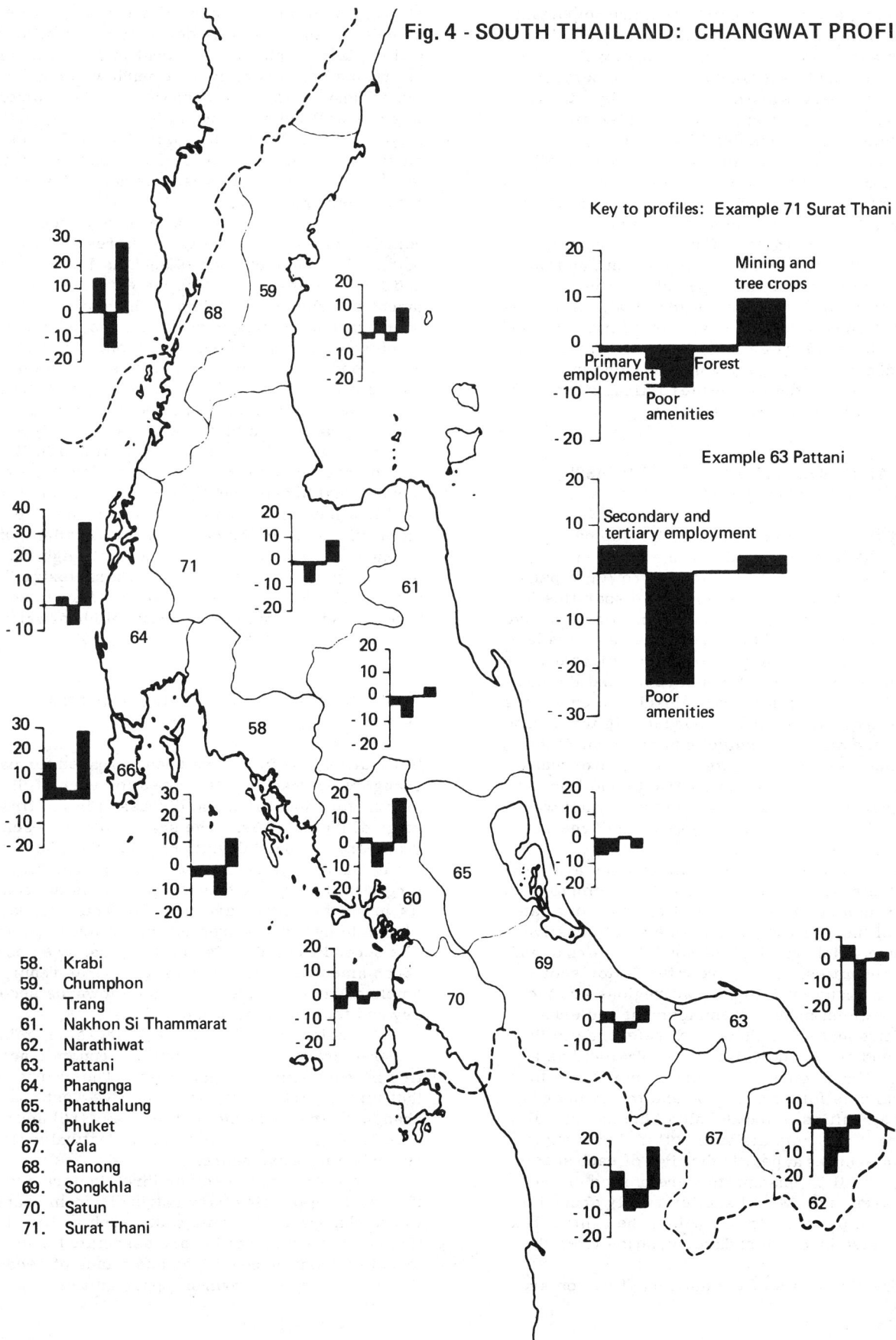

Fig. 4 - SOUTH THAILAND: CHANGWAT PROFILES 1970

Key to profiles: Example 71 Surat Thani

Mining and tree crops

Primary employment

Forest

Poor amenities

Example 63 Pattani

Secondary and tertiary employment

Poor amenities

58. Krabi
59. Chumphon
60. Trang
61. Nakhon Si Thammarat
62. Narathiwat
63. Pattani
64. Phangnga
65. Phatthalung
66. Phuket
67. Yala
68. Ranong
69. Songkhla
70. Satun
71. Surat Thani

other cases such as Narathiwat (62), Yala (67) and Songkhla (69), all adjacent to Malaysia, the economy appears to be more diversified, but the urban amenities are still poor. It is at this point of the discussion that planners with local knowledge will wish to look critically at the validity of the results and the implications they may have for planning in this particular region.

AN ALTERNATIVE METHOD OF ANALYSIS

The method of analysis discussed above works from the intercorrelations between indicators and highlights the fact that some indicators of economy, environment and social welfare are closely interrelated while others are relatively independent. Some of the factors which we have identified, mapped and graphed are composites of a number of variables and in such cases ambiguities may arise in the interpretation of the factor scores. As already noted, in the case of factor 3 for example, a changwat which records a negative score may do so because it has a higher value for forest land per capita or alternatively because it has a very small amount of riceland compared with its total population. Other aspects of importance may drop out of the analysis simply because the indicators concerned do not correlate with each other. Rural and urban household amenities, for example, are of equal interest to the planners, but the intercorrelation of the indicators used is such that the one emerges as a factor which can be graphed and mapped while the other receives no explicit recognition.

As an alternative for factor analysis we can employ additive techniques where we select indicators, reduce them to standardized scores, and average them to form composite measures. In the present context we see the two approaches as complementary rather than competitive. We have commenced with factor analysis and gained special insights about the relationships involved in the Thai space economy. We now turn to additive techniques to verify some of our initial conclusions and provide additional information about aspects where the first approach proved to be of little help. We could equally well have reversed the procedure beginning with additive techniques and then moving on to factor analysis.

MEASURES OF SOCIAL WELL-BEING IN SOUTH THAILAND

In the planning context where we are charged with the task of improving levels of living and reducing inequalities we need appropriate methods of measurement. While it is impossible to devise measures that are objective and value-free we can make comparisons which are relatively simple and unambiguous. In this part of the study we select two broad areas of social concern; one which we descibe as education and employment embraces the training and utilization of the human resource in each changwat, the other is concerned with the quality of housing and household amenities which the people enjoy. The indicators used are among those listed in Appendix 1 but since their selection involves value judgements on the part of the author they are repeated here in Tables 2 and 3.

The numerical values for each measure are tabulated for the changwats of south Thailand in Table 4 and plotted in Figure 5. The scattergram presentation is directly comparable to that in Figure 3 since each indicator has been standardized with reference to the mean and standard deviation for all 71 Thai changwats and the position on the graph thus relates the level of living in each changwat to levels of living in Thailand as a whole. Changwat Trang, for example, with values of 4. 06 and - 1. 08 stands for above the norm for Thailand as a whole in education and employment but is well below the norm for household amenities.

Figure 5(a) compares changwats in terms of education and employment shown on the vertical axis and household amenities on the horizontal axis. Changwats to the top and the right are best off; those to the bottom and the left are less fortunate. The changwats of south Thailand vary widely in levels of social well-being as portrayed by these two measures. They range between the small island of Phuket, which is well above the norm for Thailand on both measures, to Narathiwat, Pattani and Phatthalung which are among the poorest changwats in the country. Two changwats, Trang and Satun, are anomalous in that they perform well on one measure but poorly on the other. In general, the scattergram suggests that levels of education and employment in south Thailand tend to scatter around the norm for Thailand while levels of household amenities tend to fall below the norm.

At no point in the discussion so far have we focused firmly on rural living conditions. In the case of household amenities we do have pertinent results and in Figure 5(b) the measure for household amenities used in the previous graph is plotted separately for urban and rural areas of each changwat using the intermediate results of Table 4. The generally low level of household amenities is again apparent and in most cases changwats which perform badly on one measure also perform badly on the other. The exceptions are Satun and Chumphon. In the case of Satun an abnormally high standardized score for one indicator (6. 86 for item 6) outweighs the other indicators and inflates the value for rural amenities.

TOWARDS AN EVALUATION

This paper should be seen as one contribution to a dialogue that will involve planners and policy-makers.

Table 2:   Indicators used for calculating the measure of education and employment

(1) Primary school attendance index
(2) Secondary school attendance index
(3) Ratio of females to males at secondary school
(4) Literacy of population aged ten years and over
(5) Ratio of employed to unemployed among experienced workers
(6) Ratio of employed to unemployed among new workers

Note: Items 5 and 6 correspond to numbers 14 and 15 in Table 1 except that the signs are reversed to give positive values for changwats with low unemployment.

Table 3: Indicators used for calculating the measures for household amenities

(a) Urban households (data for municipal areas)
   (1) Proportion of houses built of strong materials
   (2) Proportion of households with electricity for lighting
   (3) Proportion of households with piped water
   (4) Proportion of households with a car
   (5) Proportion of households with a refrigerator
(b) Rural households (data for villages)
   (6) Proportion of houses built of strong materials
   (7) Proportion of households with electricity for lighting
   (8) Proportion of households with piped water
   (9) Proportion of households with a radio

Table 4: Provinces of South Thailand: additive measures for selected aspects of human welfare

| Province | Education and employment | Household Amenities | | |
		Urban	Rural	Urban and rural
58. Krabi	-1.24	-1.20	-1.52	-1.51
59. Chumphon	1.01	0.78	-0.73	0.14
60. Trang	4.06	-0.85	-1.09	-1.08
61. Nakhon Si Thamarat	0.00	-0.91	-1.03	-1.09
62. Narathiwat	-0.88	-1.84	-1.26	-1.79
63. Pattani	-1.26	-1.53	-0.92	-1.42
64. Phangnga	-0.61	-0.19	-0.96	-0.59
65. Phatthalung	0.26	-0.35	-0.79	-0.62
66. Phuket	0.12	0.40	0.73	0.62
67. Yala	0.09	-0.47	-0.82	-0.70
68. Ranong	0.23	0.11	-0.42	-0.14
69. Songkhla	0.69	-0.45	-0.46	-0.51
70. Satun	-0.42	-0.15	2.59	1.19
71. Surat Thani	-0.28	-1.05	-0.68	-1.00

Fig. 5 - SOUTH THAILAND: CHANGWAT STRUCTURES COMPARED USING ADDITIVE MEASURES

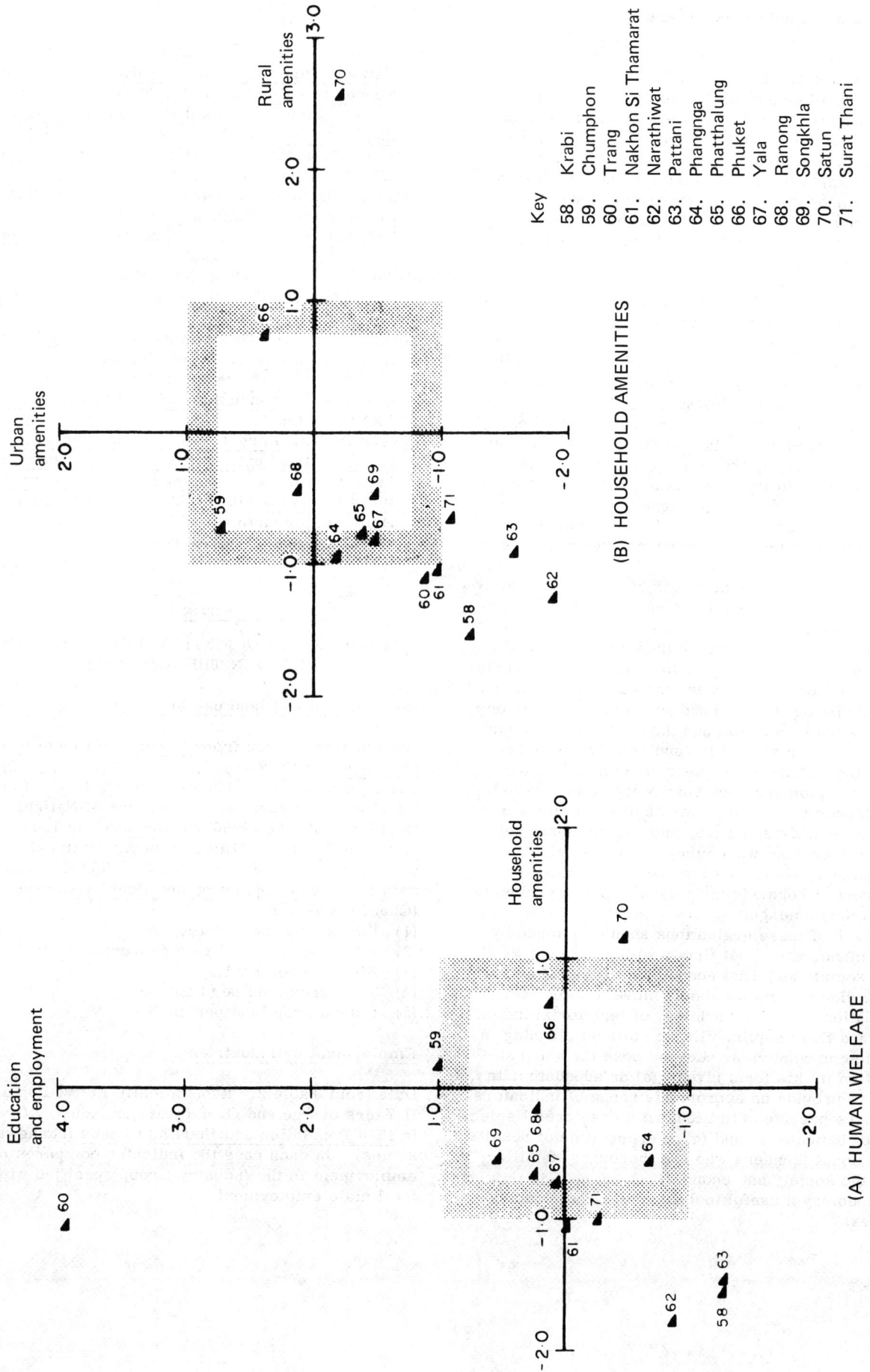

(A) HUMAN WELFARE

(B) HOUSEHOLD AMENITIES

Key

58. Krabi
59. Chumphon
60. Trang
61. Nakhon Si Thamarat
62. Narathiwat
63. Pattani
64. Phangnga
65. Phatthalung
66. Phuket
67. Yala
68. Ranong
69. Songkhla
70. Satun
71. Surat Thani

My task has been to take Thai data and to demonstrate methods of analysis and visual interpretation of results to those of you who know Thailand and are involved in the practical tasks of planning for the development of its people and its resources. From time to time in the presentation I have made some brief and tentative interpretations but these should be seen as pointers to discussion rather than attempts to arrive at substantial conclusions. The methods and the results are now open for assessment. The main value of this project will come from its evaluation by those who are familiar both with the data available and the planning tasks to be undertaken.

In our evaluations of territorial indicators as a tool for development planning, we should concentrate in turn on two different levels of discussions. To begin with you should look very carefully at the particular indicators used for the analysis and, whenever possible, suggest alternatives that are better suited to the task at hand. Questions to be answered will relate to the relevance of the indicators chosen, the validity of the definitions used and reliability of the data available. Participants may also be able to suggest sources of data for areas of concern such as health or income distribution which have not been covered in this report.

Secondly, you should look at the relevance, both of the results and the form of their presentation, to the planning task. Do the methods used help us to give a regional dimension to national planning? Do they help us to understand the structural relationships between the central region and metropolis on the one hand and peripheral regions such as the north-east and the south on the other? Do the provincial profiles and scattergrams help us to understand better the problems and resources of each region and each changwat? Can they help the planner to formulate strategies, to allocate manpower and resources, and to implement policies in localities where they are most relevant? Can an analysis of this type be used to monitor progress and provide early warning of new problems which may emerge.

Both of these evaluations should be made by Thai planners and set firmly within the context of Thai society and Thai economy. In the latter part of the discussions we should move to a broader evaluation of the methodology of territorial indicators and their applicability to national planning in developing countries. Let me pose the general question in this form given: (a) an adequate data base to provide an appropriate range of indicators, (b) more precise definition and more careful selection of indicators, and (c) interpretation of results by national planners who are thoroughly familiar with the society and economy concerned, is this methodology a useful tool for use in the planning process?

REFERENCES

Cant, R.G. (1974) The Spatial Dimensions of National Planning: The Rôle of Territorial Socio-Economic Indicators in the Formulation and Implementation of National Development Plans in the Asian Region. Methods and Analysis Division, Department of Social Sciences, Unesco, Working Paper SHC-74/WS/24.

National Economic and Social Development Board, Government of Thailand (NESDB, 1971) The Third National Economic and Social Development Plan 1972-1976, Bangkok.

National Statistical Office, Thailand (1972) 1970 Population and Housing Census (Changwat Series), Bangkok.

Royal Thai Survey Department, (1972) Thailand National Resources Atlas, Bangkok.

Unesco (1975a) "The Social Indicators Programme at Unesco" International Social Science Journal, XXVII, 1, 195-7.

Unesco (1975b) Report: Meeting of a Working Group on the Applicability of Socio-Economic Indicators to Development Planning. Bangkok, 16-24 September 1974. Methods and Analysis Division, Department of Social Sciences, Unesco, Working Paper SHC-75/WS/4.

Appendix 1

SOURCES AND DEFINITIONS OF INDICATORS
USED IN THE ANALYSIS

Environment and land use indicators

Land use data taken from Thailand National Resources Atlas, 1972 pages 9.2 to 9.5. These are based on a land classification survey by the Land Development Department, Ministry of National Development, 1961-1966 and Handbook of Land Utilization in Thailand. Division of Agricultural Economics, Ministry of Agriculture, 1965. Population data from 1970 Population and Housing Census (Changwat series).
(1) Forest land per capita
(2) Farm holding land as a proportion of total area
(3) Riceland per capita
(4) Tree crop land per capita
(5) Upland crop land per capita

Employment by industries

Data from Table 21 "Economically Active Population 11 Years of Age and Over by Major Industry Group" in 1970 Population and Housing Census (changwat series). In each case the indicator compares male employment in the Industry Group specified with total male employment.

(6) Employment in agriculture, forestry, hunting and fishing

(7) Employment in mining and quarrying

(8) Employment in manufacturing

(9) Employment in commerce, transport, storage and communication

Education and employment indicators

Data from Tables 12, 13 and 16 of 1970 Population and Housing Census (changwat series)

(10) Primary attendance index measured by percentage of children aged 12 years attending school

(11) Secondary attendance index measured by percentage of youth aged 16 years attending school

(12) Ratio of females to males among 16 year-olds attending school

(13) Per cent of population aged 10 years and over who are literate

(14) Unemployment rate among experienced male workers who are economically active

(15) Unemployment rate among new male workers who are economically active

Household amenity indicators for urban and rural areas

Data from Housing Census tables of the 1970 Population and Housing Census (Changwat series). Figures for municipal areas are used for urban indicators and include all nakhons (cities), muangs (towns) and tambons (districts). Non-municipal data are used for rural indicators. Results for non-municipal households are based on a 25 per cent sample enumeration.

(16) Urban houses built of strong materials as a proportion of all urban houses

(17) Proportion of urban households with electricity for lighting

(18) Proportion of urban households with piped water

(19) Proportion of urban households with an automobile

(20) Proportion of urban households with a refrigerator

(21) Rural houses built of strong materials as a proportion of all rural houses

(22) Proportion of rural households with piped water

(23) Proportion of rural households with a bicycle

(24) Proportion of rural households with a radio

Demographic indicators

Population data from 1970 Population and Housing Census (Changwat series). Land area of each changwat from Thailand National Resources Atlas.

(25) Population in municipal areas as a proportion of total population

(26) Population per hectare of land area

(27) Male to female ratio in total population

(28) General fertility or total number of children born alive to all women 15 years of age and over

(29) Child dependency or number of children aged 0 to 14 compared with number of adults aged 15 to 59

(30) Aged dependency or number of population aged 60 years or over compared with population aged 15 to 59

(31) Migrants who were residents elsewhere within their present changwat on 1 April 1965

(32) Migrants who were resident outside present changwat on 1 April 1965

Appendix 2: Factor scores for changwats of Thailand, 1970

	Changwat	(1) Employment structure	(2) Urban amenities	(3) Land use	(4) Tree crops and/or mining
1.	Chiang Rai	-0.58	-0.83	-0.12	-0.10
2.	Chiang Mai	0.21	-0.90	-0.20	0.41
3.	Nan	-0.52	-1.46	-0.84	-0.57
4.	Phrae	0.01	-0.37	0.11	-0.69
5.	Mae Hong Son	0.28	-3.99	-2.38	0.07
6.	Lampang	-0.09	-0.71	-0.23	-0.52
7.	Lamphun	-0.28	-0.55	-0.81	-0.70
8.	Kalasin	-0.99	-0.40	-0.10	-1.03
9.	Khon Kaen	-0.73	-0.33	0.97	-0.98
10.	Chaiyaphum	-0.98	-0.12	0.71	-0.72
11.	Nahkon Phanom	-0.55	-0.57	0.77	-1.25
12.	Nakhon Ratchasima	-0.69	-0.14	0.91	-0.48
13.	Buri Ram	-1.02	-0.47	0.15	-1.14
14.	Maha Sarakham	-0.68	-0.98	-0.91	-1.70
15.	Roi Et	-0.68	-0.36	-0.32	-1.49
16.	Loei	-0.78	-1.15	0.77	-0.26
17.	Sakon Nakhon	-0.74	-0.65	0.47	-0.98
18.	Si Sa Ket	-0.80	-0.63	-0.20	-1.39
19.	Surin	-0.86	-0.18	0.20	-1.06
20.	Nong Khai	-0.65	-0.45	1.40	-0.61
21.	Udon Thani	-0.85	-0.46	1.58	-0.64
22.	Ubon Ratchathani	-0.73	-0.28	0.02	-0.82
23.	Phra Nakhon	4.25	-0.95	1.97	-1.58
24.	Thon Buri	4.38	-0.80	1.59	-1.61
25.	Kanchanaburi	-0.54	-0.11	0.85	0.61
26.	Kamphaeng Phet	-0.96	0.09	-0.17	-0.01
27.	Chanthaburi	-0.07	0.07	0.66	1.15
28.	Chacheongsao	0.12	1.15	-0.53	-0.17
29.	Chon Buri	0.29	0.69	1.57	1.06
30.	Chai Nat	-0.30	1.47	-0.67	-0.61
31.	Trat	-0.16	-0.31	1.25	1.53
32.	Tak	-0.05	-1.41	-0.91	-0.28
33.	Nakhon Nayok	-0.15	2.03	-0.31	-0.18
34.	Nakhon Pathom	0.17	1.55	0.69	-0.01
35.	Nakhon Sawan	-0.56	1.37	0.84	-0.07

Changwat	(1) Employment structure	(2) Urban amenities	(3) Land use	(4) Tree crops and/or mining
36. Nonthaburi	1.90	-0.12	0.59	-0.52
37. Pathum Thani	0.71	1.14	-0.96	-0.45
38. Prachuap Khiri Khan	-0.47	-0.21	0.87	0.99
39. Prachin Buri	-0.11	0.68	-0.17	0.32
40. Ayutthaya	1.18	2.17	-1.52	-0.07
41. Phichit	-0.52	1.40	-0.58	-0.60
42. Phitsanulok	-0.65	0.55	0.97	-0.26
43. Phetchaburi	0.07	0.66	-0.28	-0.11
44. Phetchabun	-0.75	-0.22	0.87	0.00
45. Rayong	-0.29	0.15	1.59	0.53
46. Lop Buri	-0.32	1.13	1.16	0.53
47. Ratchaburi	0.349	0.91	0.11	0.44
48. Samut Prakan	1.94	0.46	0.73	-0.16
49. Samut Songkhram	1.73	-0.19	-1.82	-0.66
50. Samut Sakhon	1.29	0.15	-0.56	-0.72
51. Saraburi	0.03	1.30	0.06	0.31
52. Sing Buri	0.42	2.52	-0.97	-0.00
53. Sukhothai	-0.74	0.63	0.06	-0.52
54. Suphan Buri	-0.23	1.45	-1.63	0.23
55. Ang Thong	0.31	1.75	-2.12	-0.37
56. Uttaradit	-0.38	-0.32	0.35	-0.20
57. Uthai Thani	-0.66	0.94	0.52	0.09
58. Krabi	-0.46	-1.18	-0.37	1.09
59. Chumphon	-0.07	-0.20	0.58	1.01
60. Trang	0.13	-0.38	-1.00	1.81
61. Nakhon Si Thammarat	-0.36	0.04	-0.74	0.40
62. Narathiwat	0.39	-1.00	-1.88	0.52
63. Pattani	0.58	0.03	-2.36	0.35
64. Phangnga	0.04	-0.68	0.43	3.45
65. Phattalung	-0.57	0.06	-0.51	-0.32
66. Phuket	1.44	0.20	0.27	2.73
67. Yala	0.72	-0.72	-0.84	1.71
68. Ranong	0.00	-1.34	1.32	2.96
69. Songkhla	0.28	-0.18	-0.69	0.59
70. Satun	-0.46	-0.31	0.55	0.17
71. Surat Thani	-0.16	-0.14	-0.81	0.97

Appendix 3: Additive measures for (a) education and employment and
(b) household amenities in Thai changwats, 1970

Changwat	(a) Education and employment	(b) Household amenities		
		Urban	Rural	Urban and Rural
1. Chiang Rai	-0.30	-0.50	-0.91	-0.77
2. Chiang Mai	0.47	-0.14	-0.61	-0.39
3. Nan	0.12	-1.25	-1.34	-1.45
4. Phrae	0.04	0.65	-0.15	0.34
5. Mae Hong Son	-0.90	-3.20	-3.10	-3.56
6. Lampang	-0.27	0.09	-0.29	-0.08
7. Lamphun	0.20	-1.05	-0.02	-0.68
8. Kalasin	-0.05	-0.64	-0.98	-0.89
9. Khon Kaen	0.01	0.90	-0.43	0.36
10. Chaiyaphum	-0.66	0.50	-0.43	0.10
11. Nakhon Phanom	0.29	1.20	-0.13	0.70
12. Nakhon Ratchasima	-0.04	0.69	-0.43	0.23
13. Buri Ram	-0.07	-0.11	-1.09	-0.61
14. Maha Sarakham	0.90	-1.90	-1.08	-1.74
15. Roi Et	0.74	-0.47	-0.92	-0.76
16. Loei	-0.21	0.00	-0.17	-0.08
17. Sakon Nakhon	0.93	-0.05	0.05	-0.00
18. Si Sa Ket	0.19	-0.16	-1.48	-0.83
19. Surin	0.29	0.54	-1.19	-0.24
20. Nong Khai	-0.10	0.95	0.28	0.74
21. Udon Thani	0.03	1.31	-0.30	0.68
22. Ubon Ratchathani	0.66	-0.25	-0.94	-0.62
23. Phra Nakhon	1.66	2.86	1.50	2.56
24. Thon Buri	1.46	1.82	2.33	2.31
25. Kanchanaburi	0.45	0.40	0.72	0.61
26. Kamphaeng Phet	-0.72	-1.64	0.24	-0.92
27. Chanthaburi	1.06	0.62	0.48	0.63
28. Chachoengsao	0.26	0.18	-0.04	0.10
29. Chon Buri	0.34	1.08	1.34	1.35
30. Chai Nat	-0.57	-0.14	1.29	0.55
31. Trat	1.30	1.12	-0.16	0.63
32. Tak	-1.65	-1.01	1.00	-1.13
33. Nakhon Nayok	-0.62	0.79	1.01	1.00
34. Nakhon Pathom	0.15	1.77	1.25	1.74
35. Nakhon Sawan	-0.40	0.74	0.78	0.85

Changwat	(a) Education and employment	(b) Household amenities		
		Urban	Rural	Urban and Rural
36. Nonthaburi	1.33	0.79	1.77	1.38
37. Pathum Thani	-1.10	-0.42	-0.03	-0.28
38. Prachuap Khiri Khan	-0.05	-0.36	0.92	0.22
39. Prachin Buri	-0.39	-0.03	0.26	0.11
40. Ayutthaya	-2.11	0.14	0.11	0.15
41. Phichit	-0.96	-0.47	0.63	0.01
42. Phitsanulok	-0.03	0.99	0.52	0.89
43. Phetchaburi	0.61	-0.06	0.96	0.44
44. Phetchabun	-0.15	0.31	-0.22	0.09
45. Rayong	0.40	1.27	0.83	1.22
46. Lop Buri	-0.64	0.60	1.31	1.03
47. Ratchaburi	-0.19	0.52	0.61	0.63
48. Samut Prakan	-0.64	1.49	0.05	0.97
49. Samut Songkhram	2.29	-1.64	-0.35	-1.21
50. Samut Sakhon	0.54	-0.31	-0.10	-0.25
51. Saraburi	-0.50	-0.10	0.75	0.31
52. Sing Buri	-1.72	0.57	1.60	1.15
53. Sukhothai	0.06	-0.19	0.45	0.10
54. Suphan Buri	-3.46	-0.59	1.20	0.22
55. Ang Thong	-0.80	-1.52	1.00	-0.47
56. Uttaradit	0.63	0.13	0.39	0.27
57. Uthai Thani	-0.04	0.91	0.57	0.86
58. Krabi	-1.24	-1.20	-1.52	-1.51
59. Chumphon	1.01	0.78	-0.73	0.13
60. Trang	4.06	-0.85	-1.09	-1.08
61. Nakhon Si Thammarat	-0.01	-0.91	-1.03	-1.09
62. Narathiwat	-0.88	-1.84	-1.26	-1.79
63. Pattani	-1.26	-1.53	-0.92	-1.42
64. Phangnga	-0.60	-0.19	-0.96	-0.59
65. Phatthalung	0.26	-0.35	-0.79	-0.62
66. Phuket	0.12	0.40	0.73	0.62
67. Yala	0.09	-0.47	-0.82	-0.70
68. Ranong	0.23	0.11	-0.42	-0.14
69. Songkhla	0.69	-0.45	-0.46	-0.51
70. Satun	-0.42	-0.15	2.59	1.19
71. Surat Thani	-0.28	-1.05	-0.68	-1.00

Appendix 4

SELECTED REFERENCES AND APPLICATIONS OF PRINCIPAL COMPONENTS ANALYSIS AND FACTOR ANALYSIS

Berry, B. J. L. and Ray, D. M. (1966, "Multivariate Socio-Economic Regionalization: A Pilot Study in Central Canada" in Rymes T. and Ostry S. (eds.) Regional Statistical Studies, Univ. of Toronto Press, Toronto.

Blaikie, P. M. (1971), "Spatial Organisation of Agriculture in Some North Indian Villages", Transactions of the Institute of British Geographers No. 52, 1-40.

Cattell, R. B. (1965), "Factor Analysis: An introduction to essentials, I and II" Biometrics Vol. 21, 190-215 and 405-435.

Clark, D., Davies, W. K. D. and Johnston, R. J.; "The Application of Factor Analysis in Human Geography", The Statistician Vol. 23, No. 3/4, 259-281.

Cooley, W. W. and Lohnes, P. R. (1971), Multivariat Data Analysis, John Wiley and Sons, New York.

Greer-Wootten, B. (1972), A Bibliography of Statistical Applications in Geography, Association of American Geographers, Commission on College Geography Technical Paper No. 9, Washington D. C.

Guertin, W. H. and Bailey, J. P. (1970), Introduction to Modern Factor Analysis, Edwards Bros. Ann Arbor, Michigan.

Harman, H. H. (1966), Modern Factor Analysis Second Edition, Chicago University Press, Chicago, Illinois.

Hartley, R. G. and Norris, J. M. (1969), "Demographic Regions of Libya: A Principal Components Analysis of Economic and Demographic Variables", Tijdschrift voor Economische en Sociale Geografie Vol. 60, 221-227.

Horst, P. (1965), Factor Analysis of Data Matrices, Holt Rinehart and Winston, New York.

King, L. J. (1969), Statistical Analysis in Geography, Prentice Hall, Englewood Cliffs.

Lawley, D. N. and Maxwell, A. E. (1963), Factor Analysis as a Statistical Method, Butterworth and Co., London.

Mabogunje, A. L. (1965), "Urbanization in Nigeria: A Constraint on Economic Development" Economic Development and Cultural Change, Vol. 13, 413-438.

Rees, P. H. (1971), "Factorial Ecology: An Extended Definition, Survey and Critique of the Field", Economic Geography, Vol. 47, 220-233.

Rummel, R. J. (1970), Applied Factor Analysis Northwestern University Press, Chicago, Illinois.

Planning for Development in the Philippines: Case Studies to Examine the Applicability of Territorial Indicators at National and Regional Levels*

R.G. Cant

(Department of Geography, University of Canterbury, Christchurch, New Zealand)

* Unesco workshops document No. SHC-75/WS/58 dated November 1975.

PLANNING FOR DEVELOPMENT IN THE PHILIPPINES: CASE STUDIES TO EXAMINE THE APPLICABILITY OF TERRITORIAL INDICATORS AT NATIONAL AND REGIONAL LEVELS

R. G. Cant

This study is one of two pilot projects designed to explore the potential of social indicators used as a tool for use in the process of national and regional planning. In common with the report of parallel study in Thailand, it is a work-in-progress document, presented at the point where two research programmes converge. National Planners in the Philippines have widened their terms of reference and have expanded their evaluation and planning machinery to embrace social and spatial concerns. During the last two years, a comprehensive organization for regional development has been established by the National Economic and Development Authority. At the same time the Unesco programme of research on social indicators has been given a series of new thrusts one of which is designed to relate the methodology of social indicators to the practical tasks of regional planning. The material set out here is intended as a basis for dialogue between regional planners and policy makers on the one hand and research workers and academics, on the other. It runs parallel to another study designed to relate social indicators to planning at the national level (Mangahas, 1975).

The social and spatial goals of Philippine planning

The Philippines has a long and well-established tradition of national development planning. During the 1970s the terms of reference for national planning have been widened and the National Economic and Development Authority has been given new powers and new responsibilities. The Four Year Development Plan FY 1974-77, adopted in July 1973 in accordance with Presidential Proclamation No. 1157, is the first plan to take into account the social and economic changes which have been initiated since martial law was proclaimed in September 1972. The two new concerns of the plan are summed up by the Director-General Sicat in his foreword where he writes that it "provides for a wider distribution of the benefits of economic growth by placing greater emphasis on social development and by integrating the approach to regional development".

The social concerns of the Philippine Government are made much more explicit in the body of the plan. On page 16, for example, the authors look at the challenge of development in the years ahead and list a number of problems and challenges to be grappled with. Foremost among these are a high rate of population growth, unemployment and underemployment both in urban and rural areas, income inequality and an uneven distribution of the fruits of progress, low standards of living especially in rural areas and inadequate infrastructure for development.

The general objective of the plan is to improve the standard of living of the greater mass of the population. Regional development is listed among the specific objectives and its importance is implicit in the general approach to development which aims at "raising rural incomes and achieving self sufficiency in food production" and simultaneously, in the industrial sector providing a "thrust towards promotion of employment ... expansion of exports ... and strengthening of industrial linkages ... the overall effort to be supported and sustained by infrastructure development". To co-ordinate planning and development at the regional and provincial levels the National Economic and Development Authority has established a comprehensive system of Regional Councils supported by Regional Economic Directors and a full-time professional staff.

At the present time then we have a situation where the social and spatial dimensions of planning are clearly recognized and the planning staff of the National Economic Development Authority are searching for the technical means whereby they can evaluate regional needs and opportunities, identify appropriate strategies for development, and measure the progress that is being made towards the achievement of planning goals.

The social indicators programme of Unesco

The Division of Social Science Methods and Analysis* of Unesco has worked on systems of human resource indicators from 1967 onwards. In a number of theoretical and methodological studies they have used such indicators to measure levels of development and explore the interrelationships between human resources and development (Unesco, 1975, 195-6).

In 1974 two new projects were initiated; one concerned with the identification of key indicators of social and economic change, the other intended to relate the use of socio-economic indicators to the practical task of development planning and, in particular, to the problems involved in the identification and elimination of social and economic inequalities. In keeping with its intention of relating theoretical and methodological advances to practical planning a series of regional working groups were organized. One such meeting, held in Bangkok in September 1974, drew together participants from seven Asian countries, from Unesco itself, from the United Nations Institute for Economic Development and Planning in Bangkok and the Thai National Economic and Social Development Board.

The Bangkok workshop was faced with two practical questions:
"(1) Is it feasible to use socio-economic indicators for national planning?
(2) What type of indicators are appropriate?"
(Unesco, 1974, 1)
The papers and discussions at the workshop focused on the use of indicators at both the national and regional levels. One contribution by Dr. Mahar Mangahas described the system of national social indicators at present being evaluated by the Development Academy of the Philippines. Another paper by the present author examined the rôle of territorial socio-economic indicators in the formulation and implementation of national plans (Cant, 1974). The methodology discussed in that paper is now taken and applied to the Philippine situation in general and to the regional context in region 6 (West Visayas).

Task and methodology

If the planner is to devise and implement development strategies which will meet the varied needs and develop to best advantage the human and physical resources of each region and province he must be able to identify, summarize and display the pertinent characteristics of each territorial unit. He needs to be able to measure mixtures of economic activity and levels of social welfare in all parts of the Philippines and, as far as possible, be able to relate these to the natural resources and the existing infrastructure of each area. In other words, he needs to map the main features of the social and economic landscape and identify the development strengths and weaknesses of particular regions, provinces and municipalities. To carry out this task he needs three things; firstly, appropriate frameworks for spatial analysis, secondly, series of economic, social, demographic and environmental indicators and, thirdly, techniques of analysis that will enable him to identify significant relationships and map important spatial variations.

The administrative framework which has evolved in the Philippines is a particularly convenient one for spatial analyses at two different levels of resolution. There are more than sixty provinces and each of these is subdivided into municipalities. Each of the eleven planning regions contains something in excess of one hundred municipalities. It is thus possible to use provincial data for national level analysis or a regional reconnaissance and then to move to municipality level data for more detailed study of individual regions or provinces. The provinces which existed at the time of the 1970 Census are shown in Figure 1 and the planning regions established by the National Economic Development are shown in Figure 2. In the latter stages of the analysis we will examine in some detail the five provinces and 130 municipalities which make up region 6 (West Visayas).

The analysis reported on here is carried out in two stages. In section two of the report we select a battery of social indicators and use these to examine selected dimensions of well-being in the Philippines. The method of analysis demonstrated is an additive one whereby groups of indicators are defined, reduced to a standard metric by adjusting them according to the mean and standard deviation for the complete set of provinces, and then adding them together to form composite measures. These composite measures are also standardized to facilitate comparisons with other provinces and other measures. The results of the various analyses are presented in the form of maps, graphs and scattergrams.

In section three of the report we relate spatial variations in social well-being to the general planning context. This is done by incorporating other indicators which measure variations in economic, environmental or demographic conditions and using principal components analysis to examine the patterns of relationship which exist in the enlarged data matrix. A number of factors are identified in each stage of the analysis, these are scored or measured for each province or municipality and presented in the form of maps, graphs or scattergrams similar to those used in section two. In both sections the results are presented selectively to show the types of presentation which are possible and to shift the focus of attention in turn from national patterns, to the patterns in region 6 and, in some cases, to conditions in the Province of Antique within region 6.

---

* Restructured on 1 April 1976 as the Division for Socio-economic Analysis.

Fig. 1 - PHILIPPINE PROVINCES, 1970

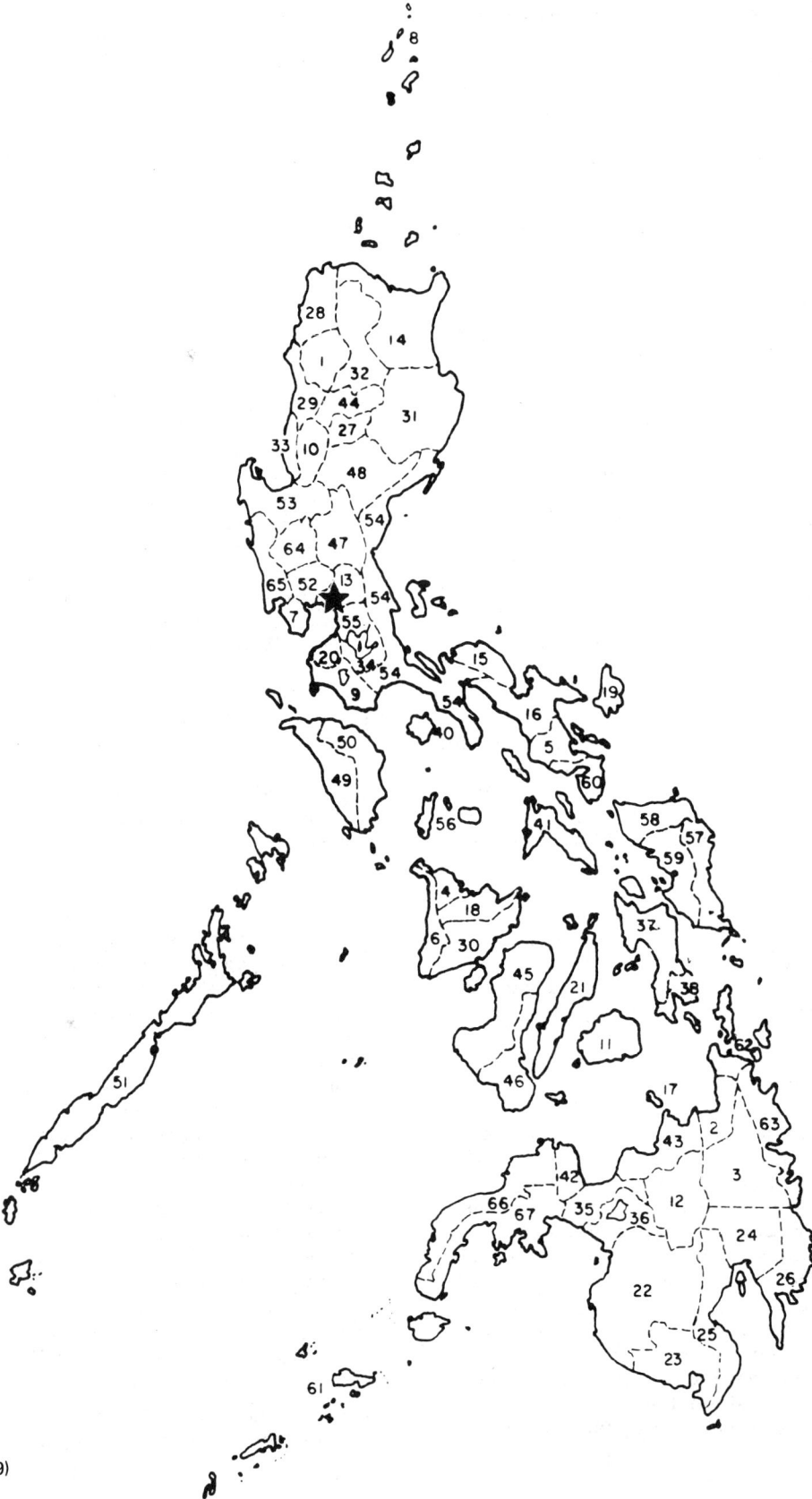

MANILA (39)

Fig. 2 - REGIONAL DELINEATION OF THE PHILIPPINES, 1974

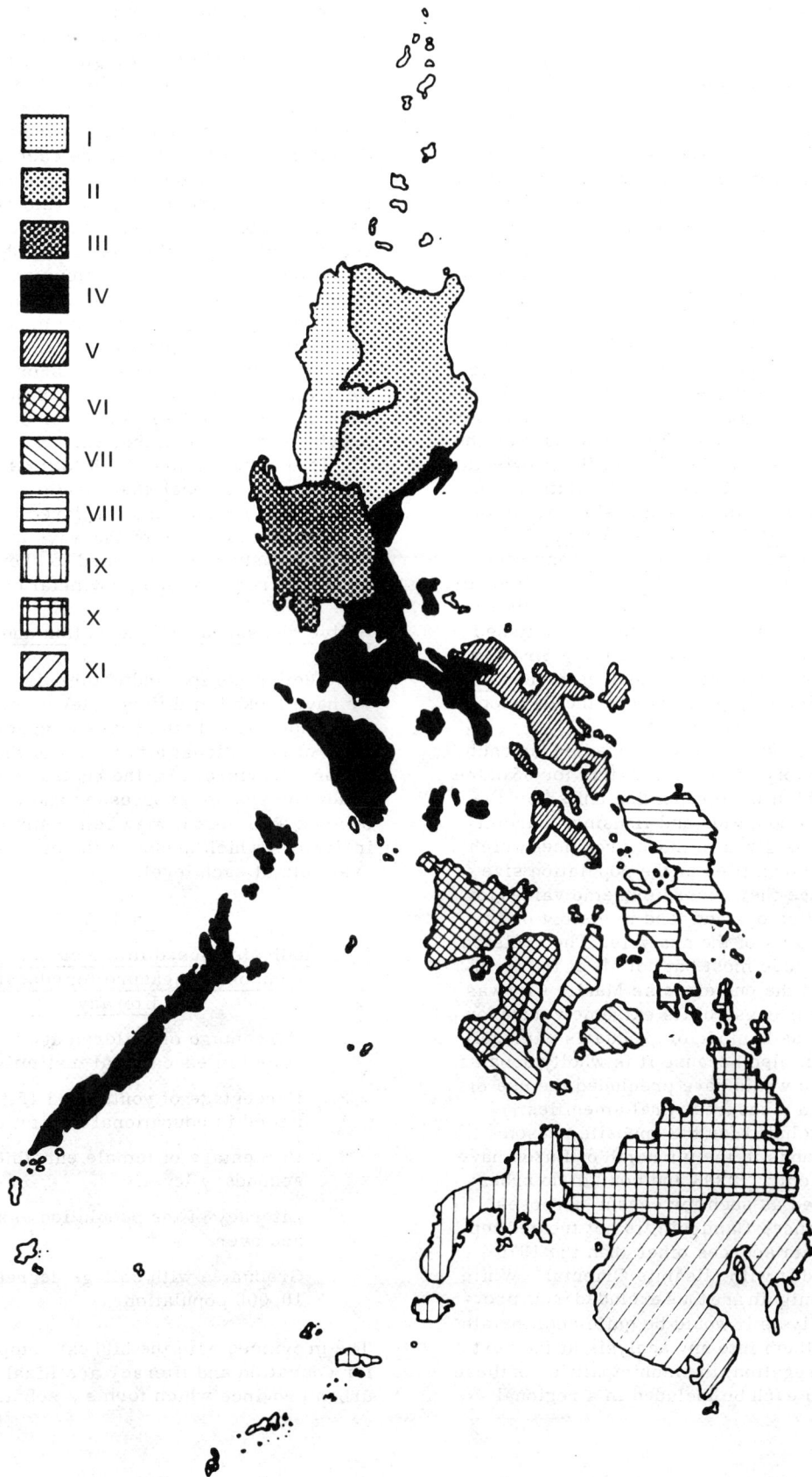

We turn first to select territorial indicators of social well-being and to answer the question "What were the spatial variations in social well-being in the Philippines in 1970?"

TERRITORIAL INDICATORS OF SOCIAL WELL-BEING

The broad dimensions of social well-being selected for territorial analysis in this section of the study represent a compromise between those which are relevant in the Philippine context and those for which recent provincial level data are available. The five dimensions considered here are: (1) education and literacy; (2) health; (3) incomes and employment; (4) housing and amenities; and (5) law and order. In some cases, those for housing, amenities and employment, data are available to provide separate measures of rural and urban conditions, in the other cases we were able to consider provincial aggregates only. More than half of the indicators used are calculated from the provincial tabulations of the 1970 Census of Population and Housing. The remaining indicators are taken or calculated from Vital Statistics Report, 1972, Indicators of Social Development, 1973 and recent numbers of the Journal of Philippine Statistics (all publications of the National Census and Statistics Office) together with Situation Reports published in 1973 as part of the Physical Planning Strategy for the Philippines and the Appendices to Human Settlements Volume 2, published by the Development Academy of the Philippines.

A battery of indicators was selected for each dimension and stored in a data matrix for 65 of the 68 provinces which were reporting units for the 1970 Census of Population and Housing. Careful consideration was given to those provinces which represent the extremities of the population size distribution since they record extreme values for a number of indicators and tend to distort or invalidate some parts of the analysis. The decision taken was to include most but not all of the territorial units. At the one extreme Manila City was excluded partly because of its extreme values for indicators such as number of graduates with college degrees but also because it is wholly urban and its inclusion would have precluded the use of indicators which measured rural amenities. Batanes was excluded for the opposite reason; these small islands to the far north of Luzon have a population of only 11,398 and contain no urban areas recognized as such for census purposes. One other province, Camiguin, is not treated separately since data sources other than the 1970 Census included it with Misamis Oriental. While Batanes and Camiguin are thus excluded from provincial level analysis it would be quite appropriate to incorporate them into any analysis at the next level of disaggregation; the municipalities of these provinces could each be included in a regional

analysis which draws on municipal level data.

Each battery of indicators is converted into a measure of well-being by means of standardized score additive model (Smith, 1973, 85-90). The signs of each indicator are adjusted so that well-being is represented by high or positive scores and lack of well-being by low or negative scores. The indicators are then rescaled to standardized scores such that the mean is zero and the standard deviation is unity. This done each battery of indicators is combined into a single measure by adding the scores which each province records for each indicator. In most cases indicators receive equal weight but in the case of number 3 where there are two income indicators and four employment indicators an adjustment is made by doubling the scores given to the income measures. Once the addition is completed the new measures are also rescaled to standardized scores. In this way we are able to relate the level of performance of each province to the level of performance for all provinces in the data matrix.

In the discussion which follows the indicators used to prepare each measure are set out and the results are mapped and displayed in various forms to demonstrate some of the ways in which the composite measures can be used to show national patterns and regional and provincial characteristics.

Active measures: (1) Education and literacy

In selecting the five indicators listed in Table 1 we have looked at different levels of instruction and emphasized both access to education and the level of educational attainment of the population in each province. As the restructuring of the education system progresses and more statistics become available it may be possible to include indicators which measure the quality of education available at each level.

Table 1

Indicators used in the calculation of a composite measure for education and literacy

1. Percentage of children aged 5 to 14 enrolled in educational institutions.

2. Percentage of youth aged 15 to 24 enrolled in educational institutions.

3. Percentage of female enrolment at the secondary level.

4. Literacy of the population aged 10 years and over.

5. Graduates with college degrees per 10,000 population.

The provinces with the highest composite scores for education and literacy are Rizal (55), a largely urban province which forms a substantial portion

Fig. 3 - COMPOSITE MEASURE FOR EDUCATION AND LITERACY

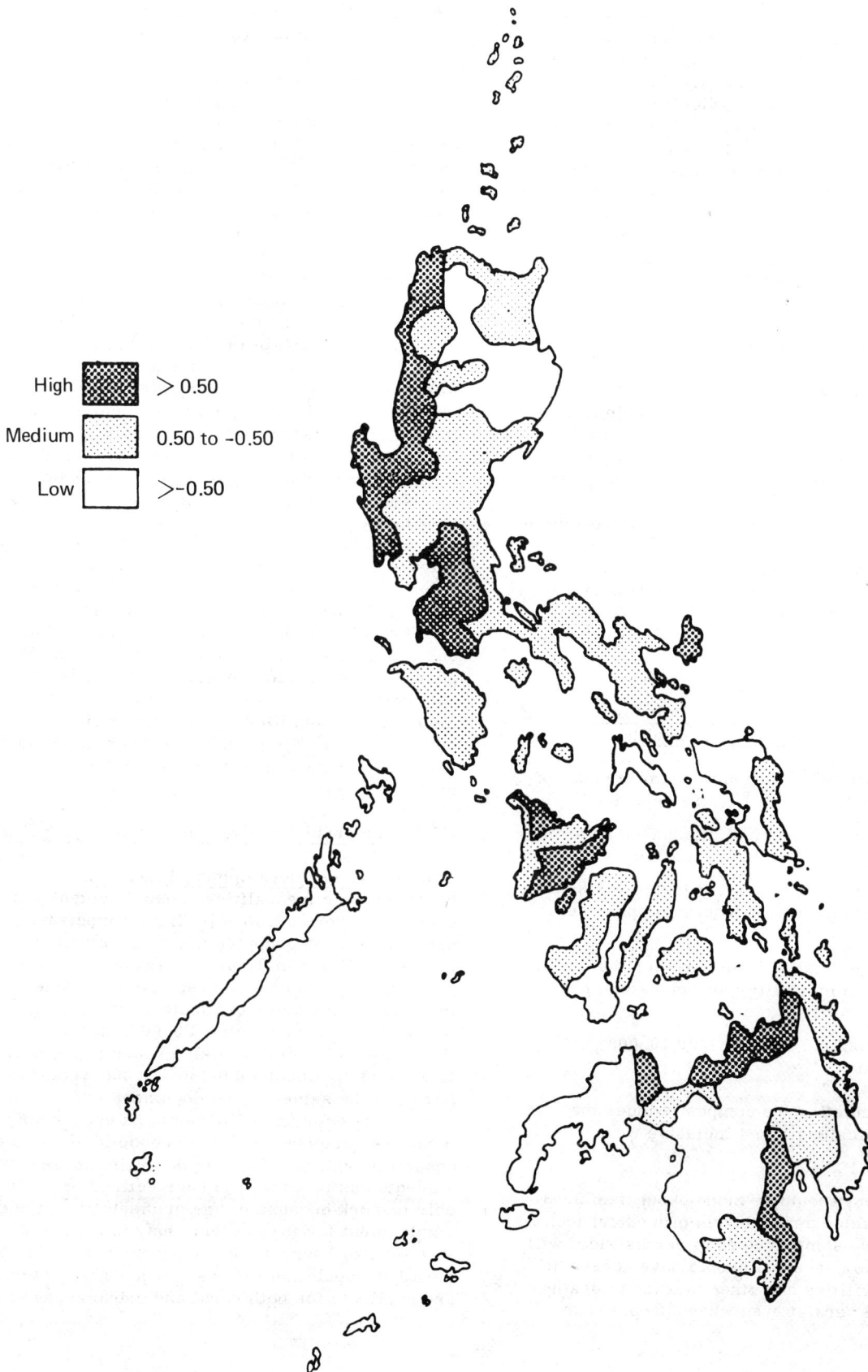

High   ▨   > 0.50

Medium   ▦   0.50 to -0.50

Low   ☐   >-0.50

of Greater Manila, and Benguet (10) which includes the resort city of Baguio. The three provinces which record the lowest scores, Lanao del Sur (36), Cotabato (22) and Sulu (61), are all in the far south and all contain a substantial proportion of Moslems. When the composite scores are mapped as in Figure 3 areas of advantage which stand out include north-west and central Luzon, Iloilo (30), northern Mindanao and Davao del Sur (25). Cebu City is also well known as an important educational centre but in this level of analysis the advantages of Cebu City are outweighed by conditions in the rural hinterland and the province as a whole falls in the intermediate category. Areas of disadvantage include most of central and western Mindanao, Sulu (61), Palawan (51), parts of Samar (58 and 59) and parts of north-east Luzon.

Additive measures: (2) Health

In selecting indicators for health we are concerned both with the health standards of the people and the level of health facilities available to them (Table 2). Where possible we have emphasized indicators which measure services relevant for the majority of the population in 1970 rather than advanced facilities which may be available only to a proportion of the population. We have thus selected items 1 and 2 which relate to medical attendance rather than hospital beds and we have set aside doctor to population ratios in favour of item 5 which compares all active medical workers with population.

Table 2

Indicators used in the calculation of a composite measure for health

1. Births without qualified medical attendance as a proportion of all births reported.

2. Deaths without medical attendance as a proportion of all deaths reported.

3. Ratio of mortality from dysentery as compared with mortality from malignant neoplasms.

4. Ratio of mortality from beri-beri as compared with mortality from malignant neoplasms.

5. Active medical workers per 10,000 population.

Note: In calculating the composite index the signs for indicators 1 to 4 inclusive were reversed.

Special comment should be made about item 3. It is argued that for existing levels of medical technology there are some classes of deaths which will be reduced to low levels if people have access to appropriate facilities and other classes of deaths which will still persist even where people have access to facilities that may rank among the best in the world. The proportion of deaths from dysentery compared with deaths from malignant neoplasms is thus suggested as an indicator which reflects levels of health and access to good medical facilities.

No data on nutrition were available at the provincial level in 1970. The closest surrogate we could find was the number of deaths from beri-beri which is built into indicator 4 similar in definition to number 3. Active medical workers, as defined for item 5, include nurses, midwives and medical technicians as well as doctors, dentists and surgeons as recorded in the occupational tabulation for the Census of Population.

The most advantaged province according to this composite measure for health is Benguet (10) which scores 2.49, ahead of Rizal with 1.88. A closer examination of the individual indicators show that Benguet has positive scores for all five whereas Rizal (55) has a very high score of 5.58 standard deviations for active medical workers (indicator 5) which is almost entirely offset by a low score of -4.42 for medical attendance at deaths. Provinces which are least advantaged in health terms are Isabela (31) and Kalinga (32) in north-east Luzon.

When composite scores for health are mapped (Figure 4) the pattern which emerges is one of relative advantage in provinces adjacent to Manila and Baguio and relative disadvantage in the case of peripheral areas such as north-east of Luzon, eastern Mindanao and parts of the Visayas. Peripheral areas with low levels on this measure are not necessarily the ones which recorded low levels for education and literacy; health services in Zamboanga del Sur, for example, appear to be much superior to educational facilities in the same province.

Additive measures: (3) Incomes and employment

The Four-Year Development Plan FY 1974-77 includes income inequalities, unemployment and underemployment among its list of problems and challenges to be grappled with. In selecting the indicators listed in Table 3 attention has been given to variables which describe areas where these problems are most acute. Indicator 2, family with incomes below P3,000, has a low threshold and is deliberately chosen in preference to a figure for incomes below P6,000 which is published in the same source document.

Figures for unemployment and underemployment are compared with the economically active population which excludes by definition housewives, students and persons who have retired or are unable to work because of age or disability. Underemployment for the present study is defined as less than 24 hours work during the week prior to the 1970 Population Census. Census tabulations are available for both rural and urban areas which

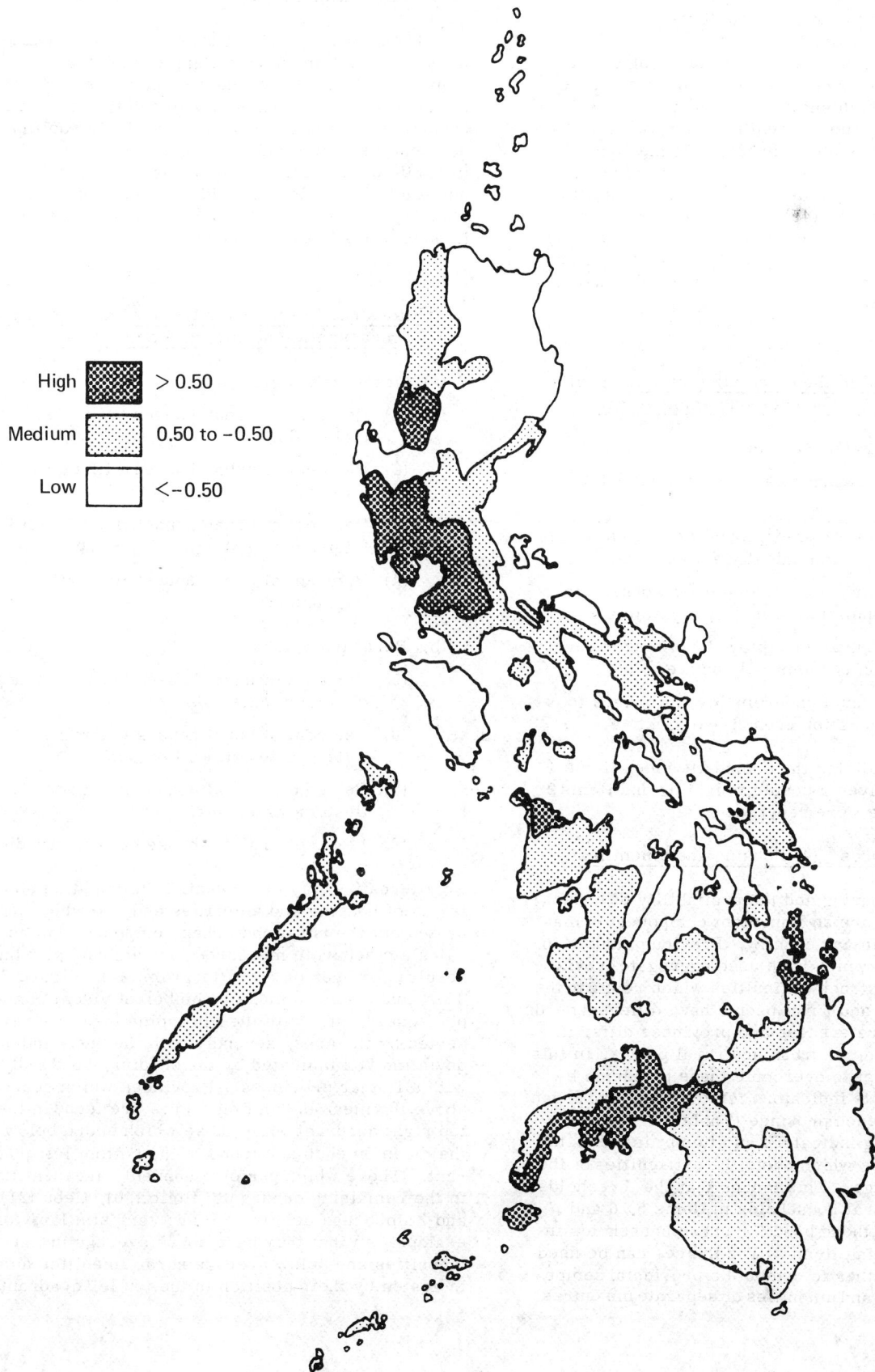

Fig. 4 - COMPOSITE MEASURE FOR HEALTH

High		> 0.50
Medium		0.50 to −0.50
Low		<−0.50

enables us to include indicators for both these areas of interest. The balance between income and employment measures is retained by adjusting the calculations to give double weight to the two income indicators.

When the composite scores for incomes and employment are mapped in Figure 5 the areas of disadvantage which show up are the provinces in the north-east and north-west of Luzon, a number of Visayan Islands and Sulu (61). At the opposite end of the scale the provinces of central Luzon adjacent to Manila, and the sugar producing province of Negros Occidental (45) are joined by the frontier provinces of Occidental Mindoro (49) and southeastern Mindanao. We will look at these results in more detail when we compare them with housing and amenities below.

Indicators used in the calculation of a composite measure for incomes and employment

1. Average family income.

2. Per cent of families with incomes below P3,000.

3. Ratio of urban unemployed workers to urban population economically active.

4. Ratio of urban underemployed workers to urban population economically active.

5. Ratio of rural unemployed workers to rural population economically active.

6. Ratio of rural underemployed workers to rural population economically active.

Note In calculating the composite index items 1 and 2 were given a double weighting and items 2 to 6 inclusive were reversed in sign.

Additive measures: (4) Housing and amenities

Standards of housing and the availability of household amenities are an important component of material welfare in any country. Previous studies in the Philippines which have used aggregate provincial data have struck difficulties which relate to the fact that urban and rural people have different needs and different preferences and provinces differ in their proportions of urban and rural people. In this study we attempt to overcome these problems by defining separate indicators for the rural and urban populations of each province (Table 4).

In selecting indicators an attempt is made to make definitions which avoid the extremities of the percentage range. This means that the thresholds selected for rural households in items 5, 6 and 7 are lower than those used for urban households in the corresponding items. The indices can be used in two ways; either to compute a provincial composite for housing and amenities or separate measures

for the rural and urban portions of each province. In this instance we have calculated all three measures and will present them in two different ways.

The broad spatial patterns for the composite measure which includes all eight indicators are mapped in Figure 6. Areas with the highest level of housing and amenities are all tightly clustered in north-west and central Luzon with the addition of Lanao del Norte (35) and Agusan del Norte (2) in north Mindanao. Provinces with low levels of household amenities are widely scattered but include a number which are located in the northern islands of the Visayan Group.

Table 4

Indicators used in the calculation of a composite measure for housing and amenities

(a) Urban indicators

(1) Per cent of urban houses built of strong materials.

(2) Per cent of urban houses with piped water.

(3) Per cent of urban households with electricity for lighting.

(4) Per cent of urban households with refrigerator.

(b) Rural indicators

(5) Per cent of rural houses built of strong or mixed materials.

(6) Per cent of rural houses with piped water, artesian well or pump.

(7) Per cent of rural households with flush or water sealed toilet.

(8) Per cent of rural households with radio.

An alternative way of presenting information about levels of housing and amenities and one which separates out the rural and urban components is to calculate separate measures for each and plot the resulting scores on a scattergram as in Figure 7. This enables us to make a number of visual comparisons; firstly with the Philippine mean represented by the axes, secondly with the one standard deviation box indicated by the shading and thirdly with all other provinces. Provinces which score above the mean on both measures are found in the top right quadrant while those which score below the mean on each are found in the bottom left quadrant. Those which perform unevenly are identified in the remaining quadrants; Iloilo (30), Cebu (21) and Zamboanga del Sur (67) are very similar, for example, in that they have above average urban amenities and below average rural amenities as indicated by their position in the top left quadrant.

Fig. 5 - COMPOSITE MEASURE FOR INCOMES AND EMPLOYMENT

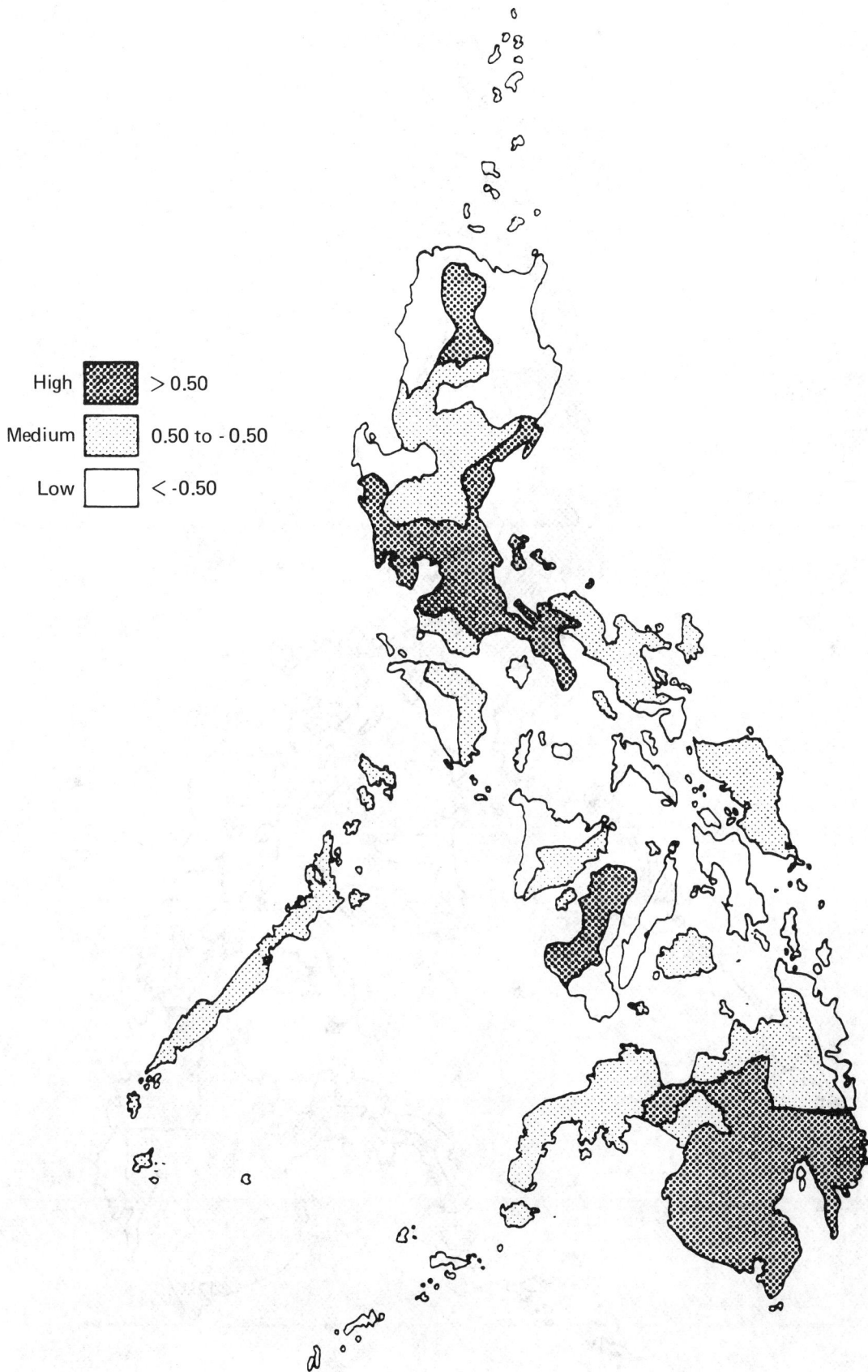

High   > 0.50

Medium   0.50 to - 0.50

Low   < -0.50

Fig. 6 - COMPOSITE MEASURE FOR HOUSING AND AMENITIES

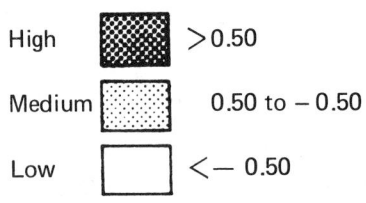

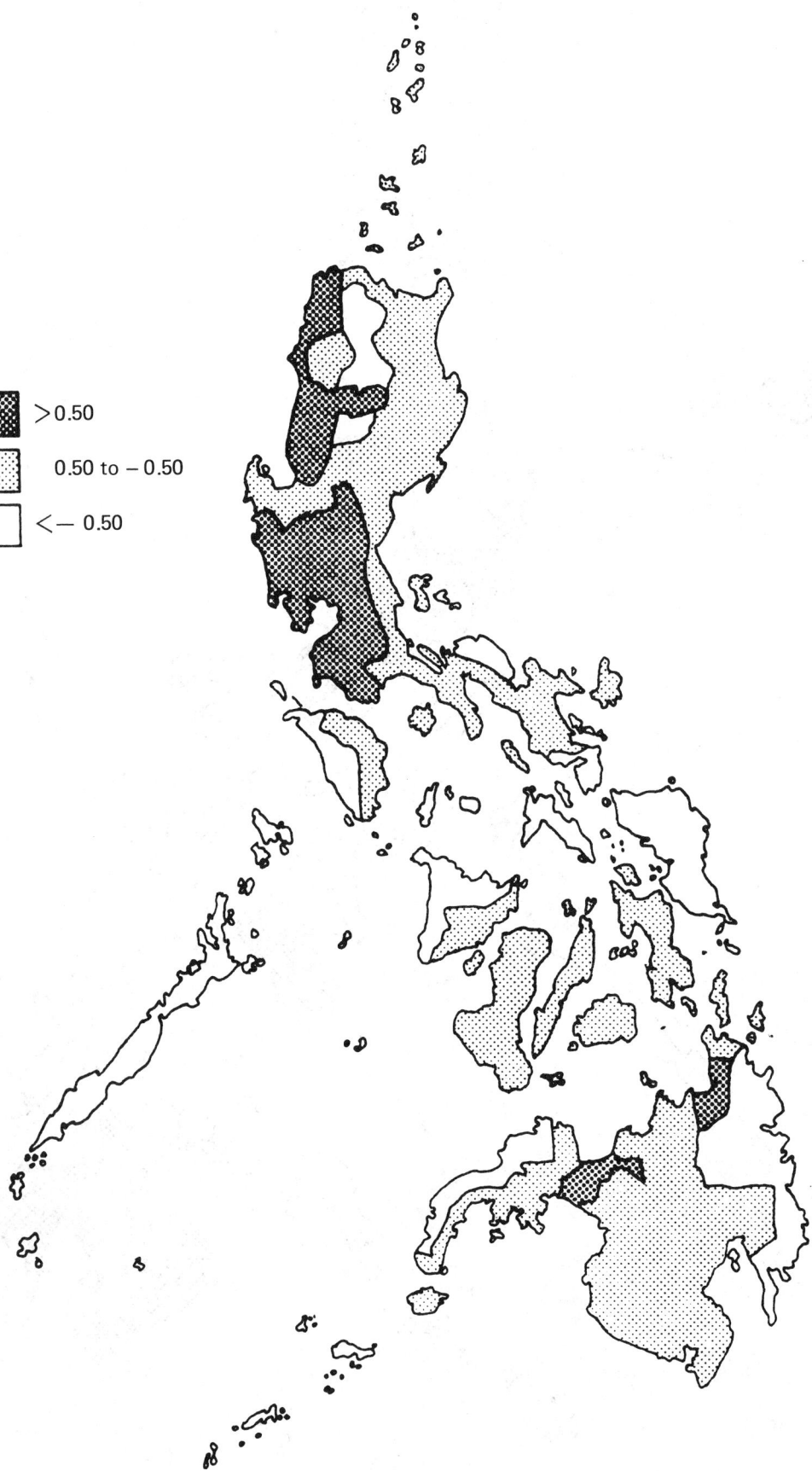

High ▨ >0.50

Medium ▨ 0.50 to − 0.50

Low ☐ < − 0.50

Fig. 7 - PHILIPPINES PROVINCES: HOUSING AND AMENITIES FOR URBAN
AND RURAL AREAS PLOTTED FROM COMPOSITE MEASURES

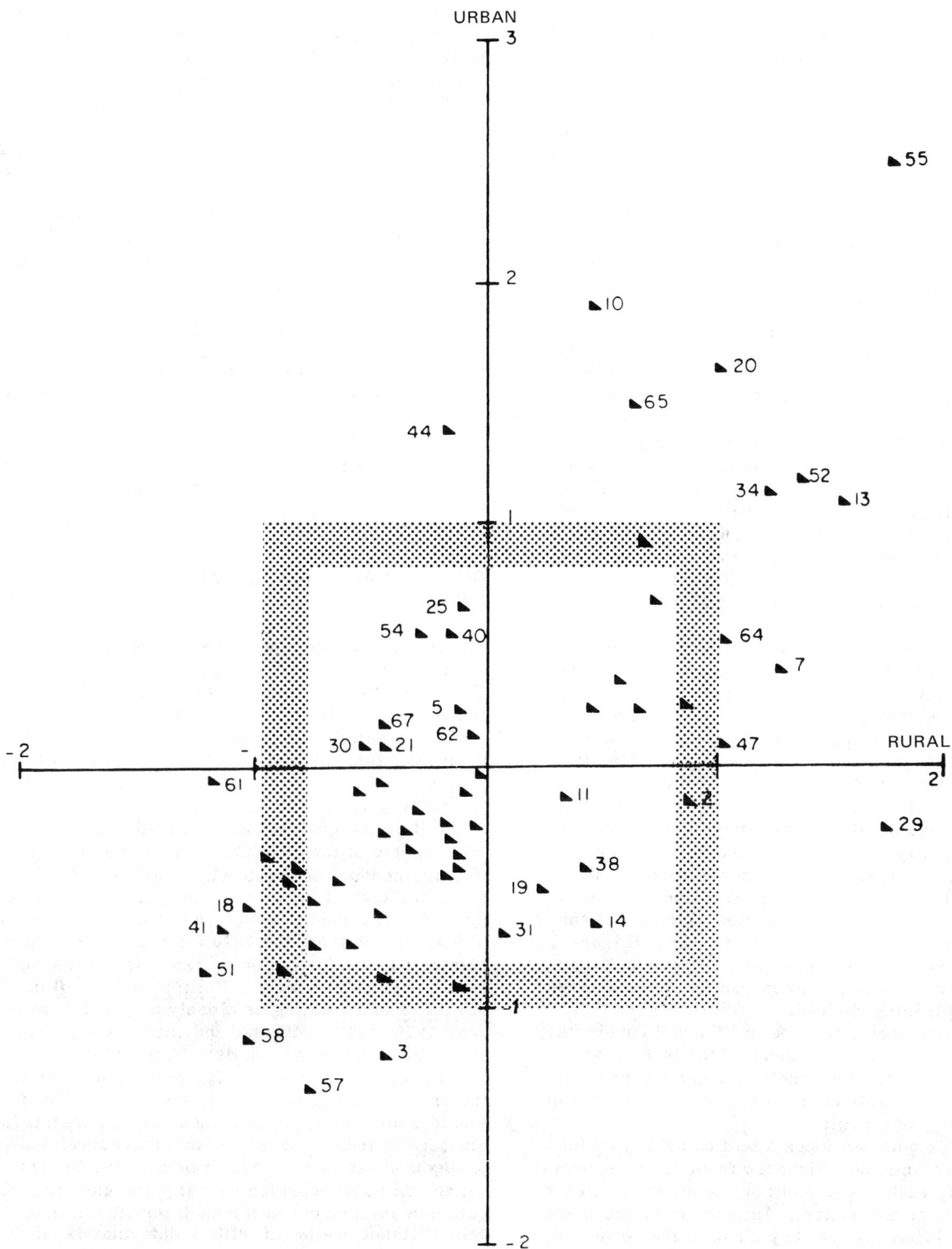

In the opposite quadrant Ilocos Sur (29) has exceptionally good rural amenities but below average urban amenities.

Additive measures: (5) Law and order

Along with the dimensions described above measures of personal security and personal freedom must be given substantial recognition. In this area, however, it is difficult to define appropriate indicators and even more difficult to find up-to-date and reliable data. Three indicators were considered; firstly the crime rate per 10,000 population; secondly a case disposal rate for the Court of First Instance which was defined as the ratio of cases disposed of during the year compared with the number of cases waiting at the start of the year; thirdly the number of inhabitants per policeman. In the first instance the indicator was abandoned because the coverage was incomplete - provinces with a law and order problem were less likely to complete the relevant statistical reports and recorded very low crime rates in the annual reports. The second indicator is still considered to be a valid measure of one aspect of the quality of justice, but the third indicator is now regarded as ambiguous in that a ratio of police to population which is adequate in a situation where law and order is already established may be quite inadequate in a situation where crime is a longstanding problem. The area of law and order is an important one and there is a need for continued research and better statistical reporting.

Composite measures and provincial profiles

We have looked in turn at five different dimensions of social well-being and in four cases we have been able to derive composite measures. The tendency in some studies has been to take a number of separate measures and synthesize them into a single composite index of social well-being. The contention made here is that such an exercise is not appropriate. Instead we would argue that it is better to focus attention on the diversity of well-being that exists in the regions and provinces of the Philippines. At this stage of the discussion we present the composite measures in the form of provincial profiles for the region which is the subject of special discussion in the workshop (Figure 8). The full results are contained in Appendix 1 and similar profiles can be prepared for the provinces of other planning regions.

The provincial profiles in Figure 8 enable us to compare the five provinces of the West Visayan Region with each other and at the same time relate their levels of well-being to levels of well-being in the country as a whole.

At this point we focus attention briefly on the profile for Antique. Planners from the other provinces may wish to comment on the other profiles in the course of discussion. Antique, in terms of education and literacy, is very close to the norm for the Philippines as a whole but is well behind the provinces of Iloilo, Aklan and Negros Occidental within region 6. It does, however, perform better in terms of health, where it stands close to Iloilo and only slightly behind Aklan. Along with Aklan it suffers serious disadvantage in the area of incomes and employment and in terms of housing and amenities it is the least fortunate of the provinces in the West Visayan Region. A major decision to be made by national planners with responsibility for provinces such as Antique, Aklan and Capiz (all of which show similar profiles) is to determine whether continued outmigration should be encouraged or whether substantial efforts should be made to increase employment opportunities and improve the quality of housing. Iloilo and Negros Occidental, each with very different profiles, obviously call for development strategies that will be quite different from each other and from those determined for the three provinces previously mentioned.

In the discussion so far attention has been concentrated on social well-being. Since social well-being does not exist in isolation from other phenomena and since planners must tackle social problems within the context of economic goals we must set these social dimensions in a broader frame of reference. In the third section of the study we examine the interrelationships between social, economic, demographic and environmental indicators.

SPATIAL VARIATIONS IN ECONOMY AND WELFARE

At this stage of the analysis we widen our terms of reference and attempt to relate variations in social well-being to variations in population, environment and economy. To do this we take a selection of the social indicators discussed above and supplement them by sets of indicators which describe the wider context. As before, the particular balance of indicators selected is influenced by the data available; at the time the analysis was carried out the provincial reports of the 1970 Census of Population and Housing had just been published, but the 1971 Census of Agriculture results were not available. Most of the variables used to describe economy and population are taken from the 1970 Census. They are supplemented by information on land use, roads and vehicles gleaned from the Philippine Journal of Statistics and relating as closely as possible to the year 1970. The additional indicators are grouped under three headings and defined in Table 5.

In contrast to the additive techniques used in the previous stage of the analysis we now demonstrate a method which searches the data matrix for clusters of indicators which are intercorrelated and presents these in the form of orthogonal factors which can be interpreted verbally and used to compute new factor scores for each territorial unit. In this instance we begin with a data matrix of 40

Fig. 8 - REGION 6. PROVINCIAL PROFILES
OF SOCIAL WELL-BEING (West Visayas)

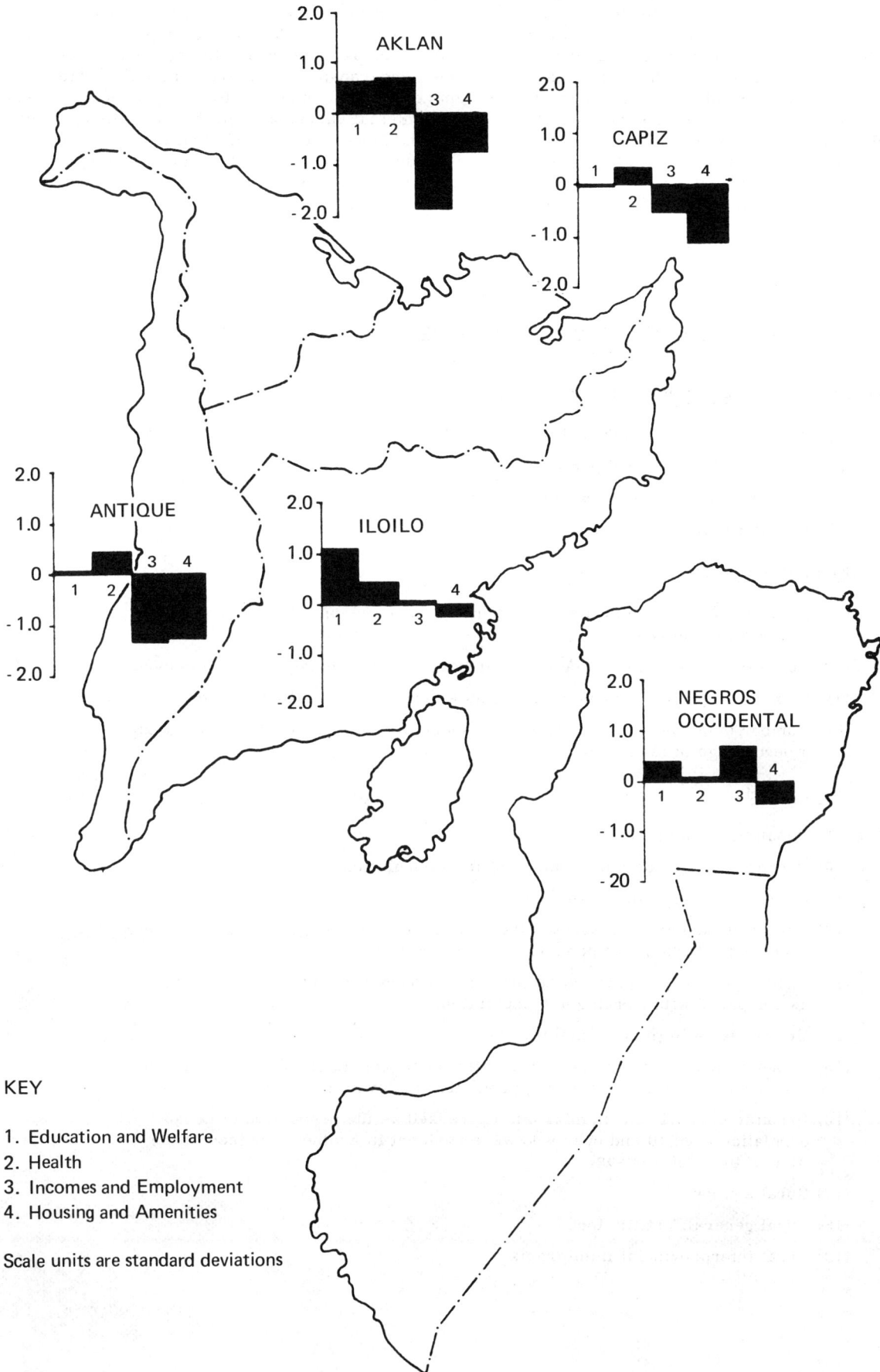

KEY

1. Education and Welfare
2. Health
3. Incomes and Employment
4. Housing and Amenities

Scale units are standard deviations

indicators recorded for 64 provinces and use a principal components analysis to replace the original indicators by a smaller number of factors which summarize the main relationships in the original matrix. In this case a number of analyses were made with numbers of factors ranging from 3 to 8 before we selected one with 3 factors as most appropriate to the task at hand. The factor loadings for this three factor solution are shown in Table 6 and the standardized factor scores which correspond to these three factors are mapped in Figure 9. In this way we are able to identify and interpret some of the broad spatial variations in economy and welfare in the Philippines as they existed in 1970.

One important point should be made before we proceed with the interpretation. In specifying and selecting the particular indicators used in this stage of the analysis some subjective decisions have been made which will influence the form of the results. The fact that there are 11 demographic indicators as compared with only four for environment and transport infrastructure will tend to focus more attention on population variations than on environmental variations. In the case of employment by industries there are only four variables involved but since each is divided by total employment a correlation has been built in which will tend to emphasize variations in employment structure.

Table 5

Additional indicators used in the provincial-level territorial analysis

(a) Environment and transport infrastructure

  (1) Area of timberland per capita.

  (2) Area of cultivated land per capita.

  (3) Road density in kilometres per hectare.

  (4) Registered motor vehicles per 10,000 population.

(b) Employment by industries

  (5) Number employed in agriculture, hunting, forestry and fishing as a percentage of all employed.

  (6) Number employed in mining and quarrying as a percentage of all employed.

  (7) Number employed in manufacturing as a percentage of all employed.

  (8) Number employed in commerce, transport, storage and communication as a percentage of all employed.

(c) Demographic

  (9) Population density.

  (10) Urban population as a percentage of total population.

  (11) Rate of population increase 1960 to 1970.

  (12) Child dependency expressed as the ratio of population aged 0 to 14 years compared with working age population, 15 to 59 years.

  (13) Aged dependency expressed as the ratio of population aged 60 years and over as compared with working age population.

  (14) Urban sex ratio (male/female).

  (15) Urban general fertility index calculated as aggregate "number of children ever born" divided by reporting population of "ever married women".

  (16) Urban interprovincial immigrants expressed as the proportion of urban population aged 10 and over who were resident in another province at the time of the 1960 Census.

  (17) Rural sex ratio.

  (18) Rural general fertility index.

  (19) Rural interprovincial immigrants.

Fig. 9 - PHILIPPINES: SPATIAL PATTERNS OF ECONOMY AND WELFARE, 1970

9C Welfare and Fertility

Low Welfare/High Fertility

■ > 0·5 S D
▨ 0·5 to −0·5
☐ < −0·5 S D

High Welfare/Low Fertility

9B Migration

Destination Areas

■ > 0·5 S D
▨ 0·5 to −0·5
☐ < −0·5 S D

Source Areas

9A Employment Structure

Secondary and Tertiary

■ > 0·5 S D
▨ 0·5 to −0·5
☐ < −0·5 S D

Primary

Table 6

The Philippines: spatial dimensions of economy and welfare, 1970[1]

Variable	Factor 1	Factor 2	Factor 3
Environment and transport infrastructure			
1. Timberland per capita	-.48	.50	
2. Cultivated land per capita	-.34		
3. Road density (km/ha)			
4. Vehicles per capita	.76		
Employment by industries			
5. Agriculture, forestry and fishing (%)	-.96		
6. Mining and quarrying (%)		-.39	
7. Manufacturing (%)	.75		
8. Commerce and transport (%)	.92		
Incomes and employment			
9. Family incomes below P3,000 (%)	-.54	-.46	
10. Urban unemployed (%)		-.51	.54
11. Urban underemployed (%)			.53
12. Rural unemployed (%)	.41	-.38	.43
13. Rural underemployed (%)		-.30	.50
Health			
14. Births without qualified attendance (%)	-.39		.32
15. Deaths from dysentery	-.57	.35	
16. Deaths from beri-beri	-.33		
17. Active medical workers per capita	.67		-.32
Education and literacy			
18. Children (5-14) enrolled (%)	.70		
19. Youth (15-24) enrolled (%)	.46		-.53
20. Females at secondary level (%)	-.50		
21. Literacy of population aged 10 plus (%)	.70		.30
Housing and amenities, urban and rural			
22. Urban houses; strong materials	.54		-.62
23. Urban households; piped water		-.45	-.38
24. Urban households; electricity	.80		-.41
25. Urban households; refrigerator	.83		-.41
26. Rural houses, strong or mixed materials	.48		-.48
27. Rural households, piped or pumped water	.66		
28. Rural households with water sealed toilet	.43		
29. Rural households with radio	.83		
Demographic			
30. Population density	.75		
31. Urban population (% total pop.)	.77		
32. Growth rate 1960-1970		.87	
33. Child dependency ratio	-.52		.40
34. Aged dependency ratio		-.68	
35. Urban sex ratio (male/female)		.61	
36. Urban general fertility index			.78
37. Urban interprovincial immigrants		.83	-.30
38. Rural sex ratio (male/female)		.74	
39. Rural general fertility index			.80
40. Rural interprovincial immigrants		.90	
Variance accounted for (%)	27.7	13.6	12.0

(1) Principal components analysis with varimax rotation of three factors.
Loadings below 0.30 are not tabulated. Loadings above 0.50 are underlined.
(2) Factor 1 = Employment structure (secondary/tertiary v's primary).
Factor 2 = Migration (source v's destination).
Factor 3 = Welfare and fertility (low welfare, high fertility v's high welfare, low fertility).

The pattern of loadings, or correlations, between the factors and the individual indicators enables us to interpret each of the factors. Factor 1 in Table 6 has very high loadings on a number of indicator variables. The largest of these include a negative loading for employment in agriculture, forestry and fishing (-0.96) and positive loadings for employment in commerce and transport (0.92) and employment in manufacturing (0.75). This factor, however, does more than indicate a distinction between areas where primary employment activities are dominant and areas where secondary and tertiary employment are important. Associated with these economic variations are important differences in health, education and housing. The links between secondary and tertiary employment and urban population and population density may be obvious ones but the analysis provides additional information when it shows that they are also associated with more motor vehicles per capita, better education (but not for girls at secondary level), a higher proportion of active medical workers and generally better standards of housing and amenities.

Provinces where primary employment predominates tend to be disadvantaged in these same respects. In the area of health, for example, fewer births are attended by qualified medical workers and a larger proportion of people die from dysentery or beri-beri. Such provinces also tend to have a higher proportion of families with low incomes and a higher proportion of child dependants. In terms of rural housing and amenities the provinces which have the lowest levels are those where there are few opportunities for employment outside the primary sector. The higher proportion of females attending secondary school is in itself an advantage but it is a characteristic which may arise in part from a lack of adequate employment opportunities for female workers.

When the factor scores for each province are mapped for factor 1 a very clear core and periphery pattern emerges. Areas with the highest levels of employment diversification and welfare are almost all located in central Luzon in a very tight cluster of provinces within or adjacent to metropolitan Manila. Areas with a narrower range of employment opportunities and lower levels of living are more dispersed but tend to be located either in interior north Luzon or in the eastern and south-western extremities of the country.

Following the specifications of the method of analysis used successive factors identify further clusters of intercorrelation which are independent of those already described. The second factor, in describing patterns of variation other than those associated with factor 1, loads heavily on a group of demographic variables. These include the growth rate of population, 1960 to 1970, and the proportions of both rural and urban inter-provincial

migrants. Factor 2 thus enables us to identify and map the areas which have been the source areas and the destinations for inter-provincial migration during the decade 1960 to 1970. Even with out-migration, such areas as north-east Luzon and Visayan Islands such as Panay (4, 6, 18, 30), Romblon (56), Bohol (11) and Leyte (37, 38) are still left with high rates of unemployment as well as a more aged population structure. Destination areas include Rizal (55) and such frontier areas as south-eastern Mindanao, interior Luzon, Occidental Mindoro (49) and Palawan (51).

The highest loadings for factor 3 are on the rural and urban general fertility indices (0.80 and 0.78). It is interesting to compare the loadings on social welfare indicators for factors 1 and 3 and to find that high fertility is most directly related to poor housing, unemployment and underemployment and is less closely related to either health or education. Provinces which perform badly in terms of this factor, those with high fertility and low levels of employment and household amenities, cluster together in north-eastern Luzon, the Bicol Region, Mindoro (49, 50), Romblon (56), Samar (57, 58) and Southern Leyte (38). In region 6 the province of Antique (6) shows the same characteristics. If the implications of this factor are substantiated by further research these would be priority areas for development policies which aim to slow the population advance by improving levels of housing and employment in both urban and rural areas.

Towards a regional level of analysis

Factors 1 and 3 together cover important areas of economy and welfare. In Figure 10 we bring these two factors together in the form of scattergrams which plot the performance of each province and relate it to the mean and standard deviation for the Philippines as a whole. For convenience the provinces and regions are separated into three broad groupings - Luzon and adjacent islands, the Visayan Islands and Mindanao-Sulu. Apart from Iloilo (30) and Cebu (21), both of which contain large commercial cities, and the resource rich province of Negros Occidental (45), the Visayan Provinces are at a relative disadvantage in terms of both factors (Figure 10b). One of the major challenges faced by national and regional planners is to devise development strategies which will improve conditions in such regions as the East Visayas, made up of the islands of Samar (57, 58, 59) and Leyte (37, 38), and such provinces as Antique (6).

When all three factors are taken together and used to plot provincial profiles for each of the provinces in the West Visayan Region we have an overview of the planning task (Figure 11). In Aklan, Antique and Capiz the planning context appears to be broadly similar; all three lack employment opportunities at the secondary and tertiary levels, all three are caught up in a circular situation where poor housing, insufficient employment and high

Fig. 10 - PHILIPPINE PROVINCES: EMPLOYMENT STRUCTURE AND
WELFARE/FERTILITY COMPARED BY PLOTTING FACTOR SCORES

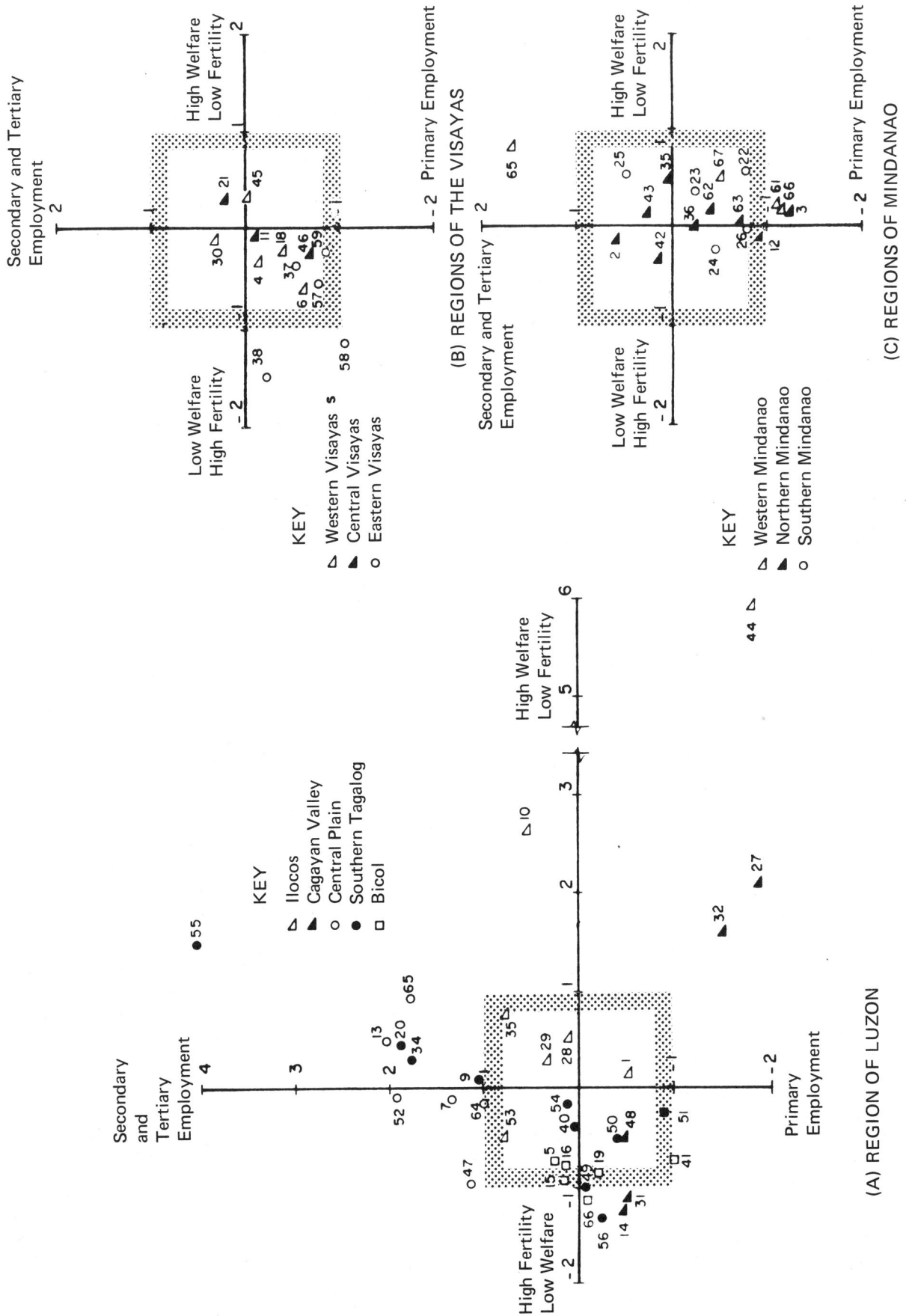

(A) REGION OF LUZON

(B) REGIONS OF THE VISAYAS

(C) REGIONS OF MINDANAO

Fig. 11 - REGION 6. PROVINCIAL PROFILES, 1970

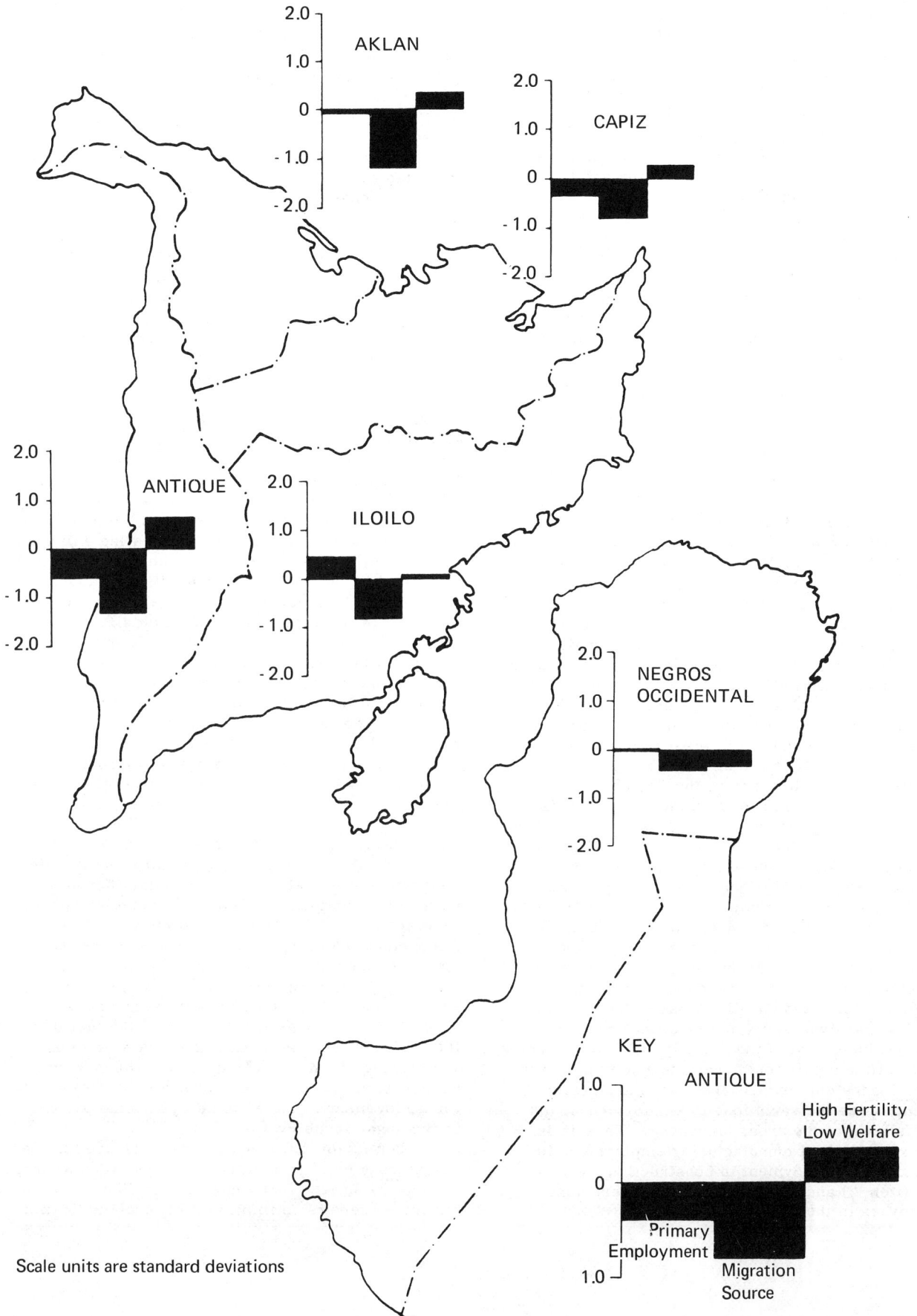

Scale units are standard deviations

fertility sustain each other and there is, as a result, substantial outmigration. Iloilo, in contrast, has a wider range of economic opportunities, but the fact that there is still substantial outmigration suggests that there may be parts of the province where the economic opportunities do not match the human resources available. Negros Occidental, similarly, is a large and varied province; the provincial profile shows it to be close to the Philippine mean on employment structure, to be losing some population from migration and to be more fortunate than its West Visayan neighbours as far as welfare and fertility are concerned. To test these initial impressions, however, we should shift our level of analysis from the broad provincial data matrix to the more detailed level of the municipality. In doing this we will forfeit a number of useful indicators, but we will gain a better understanding of the spatial variations within the West Visayan Region.

The West Visayan Region and the municipality level data matrix

In this final stage of the analysis we turn to one of the planning regions and use municipality level data to provide more detailed locational information about the dimensions we have discussed above. The West Visayan Region (region 6) is made up of five provinces which together contain a total of 130 municipalities. The Provincial Reports of the 1970 Census of Population and Housing provide detailed information about each of these small territorial units. When the project was initiated it was hoped to supplement this with additional information drawn from regional or provincial sources. This, however, proved to be impracticable; some information was available for all municipalities of each province but differences in definitions used in data collection or differences in the date of collection made it impossible to obtain comparable information for all municipalities in the Region.

The data matrix and the method of analysis used in this stage are similar to the one discussed above. This time there are 130 territorial units and 24 indicators. These indicators are grouped together under the five headings of manpower and employment, unemployment, education and literacy, demography and housing and amenities (Table 7). For reasons discussed above there are no indicators for health, environment or physical infrastructure. In general the indicators selected and the definitions used parallel those of the provincial level analysis but some modifications are made. Since most of the municipalities are rural areas with a small commercial and administrative poblacion the employment indicators are selected to emphasize the variations in employment found at this level of the urban hierarchy. Thus it is, for example, that munufacturing employment is replaced by employment in construction and repair (item 2) and the proportion of government employees in the population (item 5).

The data available can be used either to prepare composite measures of (say) education and literacy and housing and amenities or to examine the relationships between economy, welfare and demography. In this example we follow the latter course and use the principal components method applied in the previous section. A number of analyses were made involving different numbers of factors, but the one selected for presentation is a two factor solution which extends the results reported at the provincial level. This time the interest focuses less on the patterns of relationship shown in the table of factor loadings (Table 8) and more on the spatial patterns which can be mapped from the factor scores (Figures 12 and 13).

Factor 1 in Table 8 is very similar in its loadings to the first factor in the provincial level analysis (Table 6). It has a high negative loading for employment in primary industries (-0.87) and positive loadings for the cluster of service occupations which would normally be located in urban centres. Not surprisingly there are also positive loadings for the education and literacy indicators and for the housing and amenities indicators. The negative loading for child dependency indicates a higher proportion of children in the rural areas. Although we have called this factor an employment factor (thus underlining the link with the provincial level analysis), it could equally be called an urbanization factor. When we turn to the scores for each municipality on this factor and either map them (as in Figure 12) or graph them as we do for Antique in Figure 14 we have a much more detailed picture of the development strengths of each part of the region. Planners familiar with region 6 or its individual provinces will wish to comment on the validity and the implications of Figure 12.

The second factor shown in Table 7 is similar to factor 3 of the earlier discussion (Table 6). The correlations between high levels of fertility and low levels of housing and conditions of unemployment are again evident but there is a high negative loading for aged dependency (-0.84) which was not present in the earlier results. We would tentatively interpret factor 2 as an age structure factor which separates areas with an older population from areas with a more youthful population. This factor appears to be identified with employment opportunities and a movement of young people to areas where there are more jobs and better living conditions. Such an interpretation raises questions about indicator 13 which has small loadings on both factors. Was the rate of increase figure influenced by the fact that the 1960 Census was taken in February and the 1970 Census in May? It may be that long-term population movements were concealed by the short-term movements related to seasonal employment in the sugar industry.

When factor 2 is mapped, as in Figure 13, the very strong demographic contrasts between Negros Occidental and the four provinces on Panay are apparent. There is an important general implication

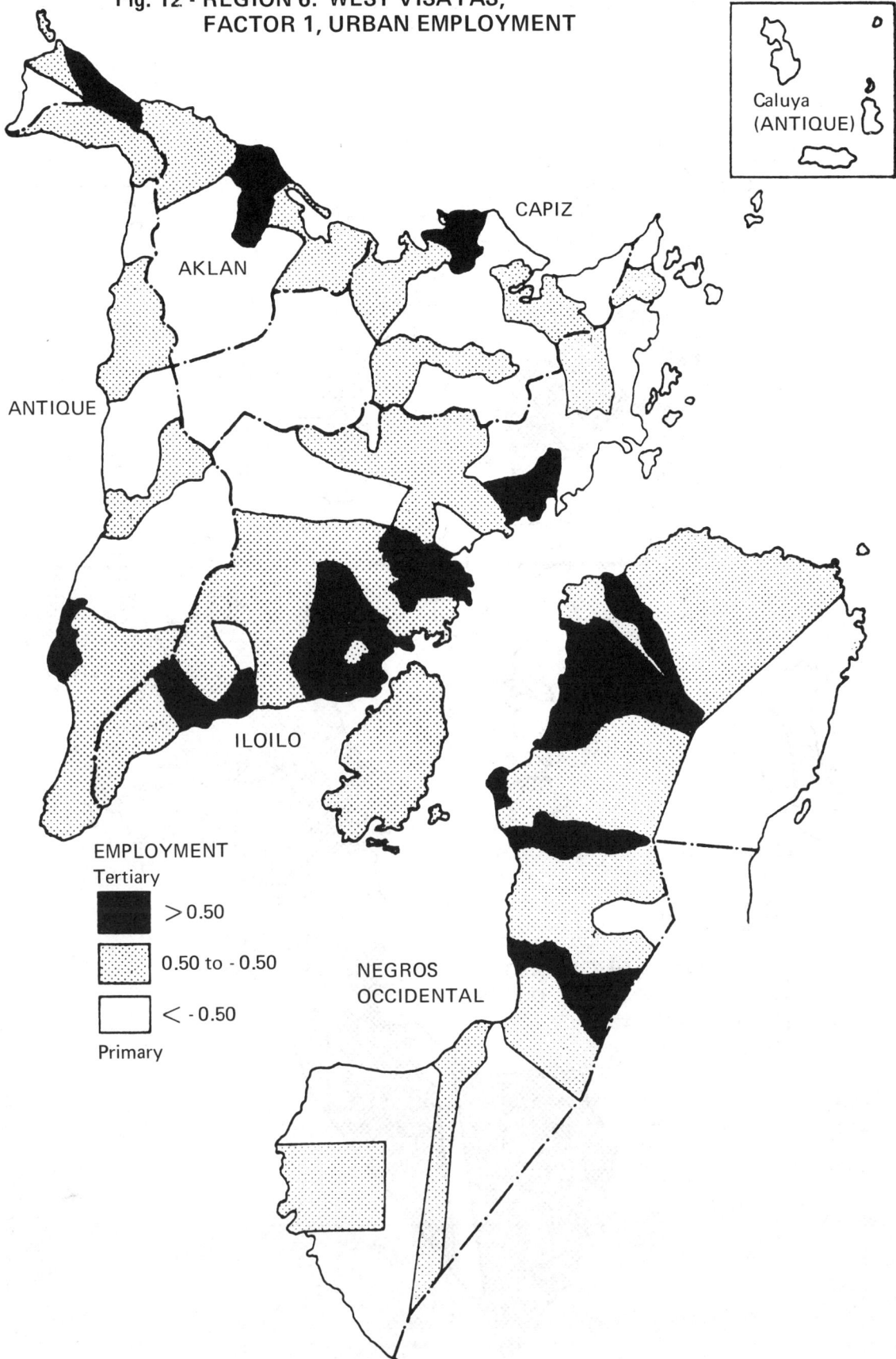

Fig. 12 - REGION 6. WEST VISAYAS,
FACTOR 1, URBAN EMPLOYMENT

Caluya
(ANTIQUE)

CAPIZ

AKLAN

ANTIQUE

ILOILO

EMPLOYMENT
Tertiary

> 0.50

0.50 to - 0.50

< - 0.50

Primary

NEGROS
OCCIDENTAL

Fig. 13 - REGION 6. WEST VISAYAS, FACTOR 2, POPULATION STRUCTURE

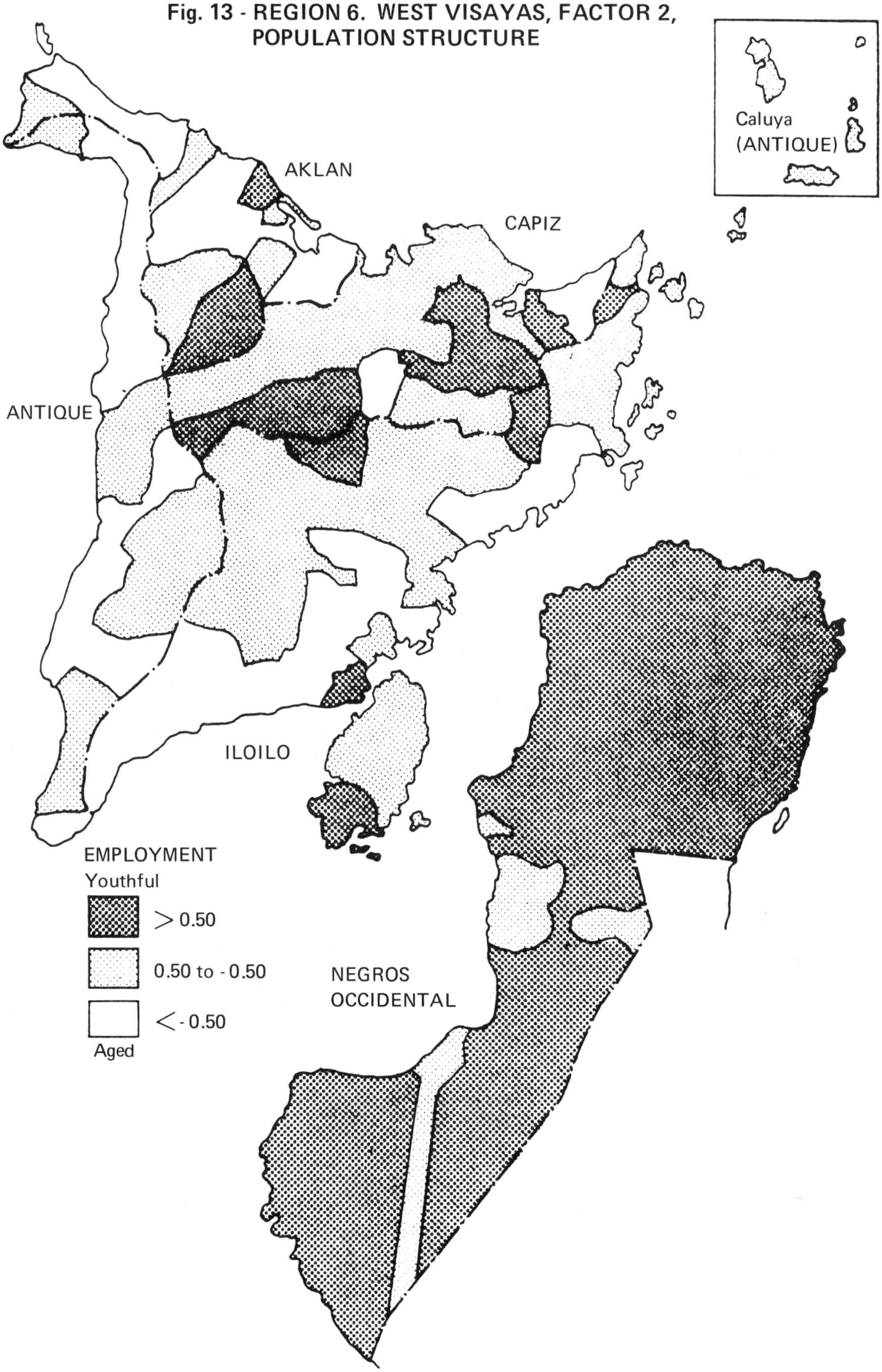

Caluya (ANTIQUE)

AKLAN

CAPIZ

ANTIQUE

ILOILO

EMPLOYMENT
Youthful

> 0.50

0.50 to - 0.50

< - 0.50
Aged

NEGROS OCCIDENTAL

Table 7

Indicators used in the analysis of economy and welfare in the West Visayas (region 6) 1970

Manpower and employment

1. Experienced workers in agriculture, forestry and fisheries as a proportion of all experienced workers.

2. Experienced workers in construction and repair as a proportion of total population.

3. Experienced workers in transport and storage as a proportion of total population.

4. Professional, technical, administrative, executive and managerial workers as a proportion of total population.

5. Government employees as a proportion of total population.

Unemployment

6. General unemployment/total population.

7. School leaver unemployment/total population.

Education and literacy

8. Primary school attendance index (percentage of 9-year olds attending school).

9. Secondary school attendance index (percentage of 15-year olds attending school).

10. Literacy of the population aged 10 years and over.

11. Population able to speak English as a proportion of population aged 10 years and over.

12. Proportion of population aged 15 years and over with 4 or more grades of college education.

Demography

13. Population growth rate 1960 to 1970.

14. Child dependency ratio (children under 15 years as a proportion of adults aged 15 to 59 years).

15. Aged dependency ratio (adults aged 60 and over as a proportion of population aged 15 to 59 years).

16. Sex ratio (male/female).

17. General fertility (aggregate number of children born alive to total number of ever-married women).

18. Within province migrants (proportion of population 10 years and over who were resident in another municipality of the same province in 1960).

19. Inter-provincial migrants (proportion of population 10 years and over who were resident in another province in 1960).

Housing and amenities

20. Strong dwellings (dwellings built of strong materials as a proportion of total dwellings).

21. Households with piped water, artesian well or pump as a proportion of all households.

22. Households with electricity as a proportion of all households.

23. Households with flush or water sealed toilets as a proportion of all households.

24. Households with a radio as a proportion of all households.

Table 8

West Visayan Region: Spatial dimensions of economy and welfare, 1970[1]

Variable	Factor 1[2]	Factor 2[2]
Manpower and employment		
1. Primary employment	-0.87	
2. Construction and repairs	0.66	
3. Transport and storage	0.56	
4. Professional and managerial	0.69	
5. Government employees	0.58	
Unemployment		
6. General unemployment		-0.62
7. School leaver unemployment		-0.47
Education and literacy		
8. Primary attendance	0.64	
9. Secondary attendance	0.55	
10. Literacy	0.76	
11. English language	0.80	
12. College education	0.85	
Demography		
13. Population growth 1960-1970		
14. Child dependency	-0.62	0.42
15. Aged dependency		-0.84
16. Sex ratio (male/female)	-0.31	
17. General fertility		-0.50
18. Local migrants		
19. Inter-provincial migrants	0.41	
Housing and amenities		
20. Strong houses	0.51	0.60
21. Water supply	0.58	
22. Electricity	0.64	0.58
23. Toilets	0.73	
24. Radios	0.64	

(1) Principal components analysis with varimax rotation of two factors.
(2) Factor 1 = Employment structure, Factor 2 = Age structure.

here - in the case of Negros Occidental there are physical resources which are relatively abundant compared with present levels of population and the first phase of modern economic development is still working through. In the case of Panay the pioneer phase is largely complete and the emphasis in planning is shifting to secondary phases of development which will use the existing resources, both human and physical, in a more efficient and intensive manner.

A focus on the province of Antique

The previous discussions have suggested that the provinces of Aklan, Antique and Capiz represent development challenges that are very different from those faced by either Negros Occidental or Iloilo. In a final figure we bring together the factor scores for the employment and age structure factors and plot them for the municipalities of the Province of Antique (Figure 14). The axis of the scattergram and the box drawn one unit from the origin in this instance represent the mean and standard deviation for Region 6. Similar scattergrams could be prepared for other provinces using the results tabulated in full in Appendix 2.

The scattergram shows that the majority of municipalities in Antique fall below the mean in terms of tertiary employment and have a population which is older than that for the West Visayas as a whole. These municipalities which are particularly limited in terms of employment and which offer a very limited range of urban facilities are Lawa-an (10), Valderrama (18) and San Remigio (14). In the latter two cases, the municipalities are both located inland away from the coastal road and may have access to facilities in adjacent municipalities. Lawan-an, by comparison, may be less fortunate.

Towards an evaluation

The material presented above is now open for evaluation by regional and provincial planners in the Philippines. It represents an attempt to select an appropriate range of social indicators and to demonstrate how two different techniques of analysis can be applied to them. The data used in the analyses have been drawn from a variety of sources and are applied at two different levels of disaggregation. The intention has been to focus attention on the same types of problem at different levels of resolution ranging from that of the national space economy, through region and province to the level of the municipality.

There has been no attempt at a comprehensive coverage dealing equally with all regions and all provinces. The aim has been to demonstrate a variety of ways in which material can be analysed and the numerical results displayed visually. At particular stages of the discussion particular interpretations have been suggested. These, however, should not be regarded as an attempt to evaluate the results in depth; rather they should be seen as markers which indicate points where individual planners will wish to examine the results and make their own interpretation. Detailed evaluation of the results for the Philippines as a whole, or for the West Visayan Region or for the individual provinces is the domain of the local planner and the local policy maker, the person with a sound grasp of the Philippine context be it national, regional, provincial or local.

It will help to organize the discussion which follows if the material is evaluated first from the perspective of the Philippine planner. Such evaluation should be made at several different levels.

(1) The indicators used. Each of the indicators should be carefully scrutinized before it is accepted or rejected for further research. Is it appropriate for the task at hand? Is the definition valid? Are the data used in its measurement accurate? Should the indicator stand alone or should it be used together with other indicators? What important dimensions are not covered in the analysis and what indicators can be devised to incorporate these?

(2) The techniques used. What are the advantages and disadvantages of the additive and correlative methods demonstrated? Which is best for the task at hand or are they best used in conjunction with each other?

(3) The results presented. In what ways can the results be used in the planning process? Are they useful in the formulation of strategies at national or regional level? Does their particular strength lie in the description of baseline conditions from which change can be measured? Will such information assist in the identification of priority areas for the implementation of particular programmes? Could territorial indicators be used to monitor the progress or provide early warning of problems that arise during the currency of a development plan?

Evaluations of this type should be made by Philippine planners and set firmly within the context of Philippine society and Philippine economy. In the latter part of the discussions we should move to a broader evaluation of the methodology of territorial indicators and their applicability to national planning in developing countries in general. The final question can be posed in this form given:

    (a) an adequate data base to provide an appropriate range of indicators;

    (b) more precise definition and more careful selection of indicators; and

    (c) interpretation of results by national planners who are thoroughly familiar with the society and economy concerned;

is the methodology of territorial indicators a useful tool for use in the planning process?

Fig. 14 - EMPLOYMENT AND AGE STRUCTURE FOR ANTIQUE PROVINCE

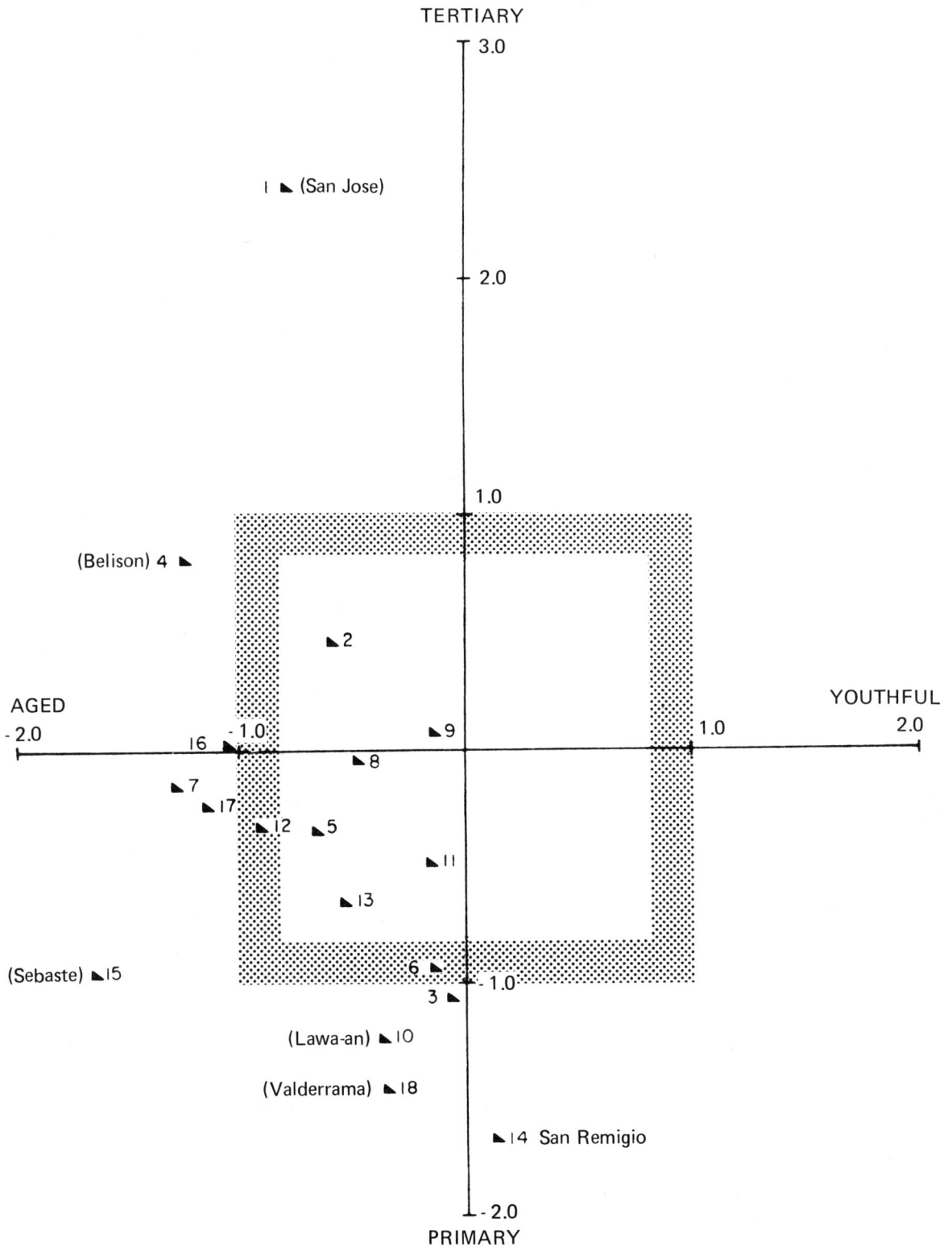

TERTIARY

3.0

I ▲ (San Jose)

2.0

1.0

(Belison) 4 ▲

▲ 2

AGED

YOUTHFUL

-2.0

16 ▲ -1.0

▲ 9

1.0

2.0

▲ 8

▲ 7

▲ 17

▲ 12 ▲ 5

▲ 11

▲ 13

(Sebaste) ▲ 15

6 ▲

-1.0

3 ▲

(Lawa-an) ▲ 10

(Valderrama) ▲ 18

▲ 14 San Remigio

-2.0

PRIMARY

REFERENCES

Bureau of the Census and Statistics (1973), Indicators of Social Development, Manila.

Cant, R.G. (1974), The Spatial Dimensions of National Planning: The Rôle of Territorial Socio-Economic Indicators in the Formulation and Implementation of National Development Plans in the Asian Region, Methods and Analysis Division, Department of Social Sciences, Unesco Working Paper SHC-74/WS/24.

Cant, R.G. (1975), Spatial Patterns and Regional Structures in Thailand: An Application of Territorial Indicators as an Input in the Development Planning Process, Methods and Analysis Division, Department of Social Sciences, Unesco Working Paper.

Development Academy of the Philippines (1974), Human Settlements: The Vision of a new Society, Report of the Task Force on Human Settlements, 4 Vols., Manila.

Development Academy of the Philippines (1975), Measuring the Quality of Life: Philippine Social Indicators, Manila.

Journal of Philippine Statistics (quarterly).

Mangahas, M. (1974), The Measurement of Philippine National Welfare, unpublished report, Manila.

Mangahas, M. (1975).

National Census and Statistics Office (1973), see Bureau of the Census and Statistics (1973).

National Census and Statistics Office (1974), 1970 Census of Population and Housing Final Report Vol. 1 (Provincial Reports), Manila.

National Census and Statistics Office (1974a), Vital Statistics Report 1972, Manila.

National Economic and Development Authority (1973), Four-Year Development Plan FY 1974-1977, Manila.

National Physical Planning Team (1973), Physical Planning Strategy for the Philippines: Situation Report, 11 Regional Reports, Manila.

Smith, D.M. (1973), The Geography of Social Well-being in the United States: An Introduction to Territorial Indicators, McGraw-Hill, New York.

Unesco (1974), Report: Meeting of a Working Group on the Applicability of Socio-Economic Indicators to Development Planning. Bangkok, 16-24 September 1974. Methods and Analysis Division, Department of Social Sciences, Unesco Working Paper SHC-75/WS/4.

Unesco (1975) "The Social Indicators programme at Unesco" International Social Science Journal, XXVII, 1, 195-197.

Appendix 1

COMPOSITE SCORES FOR SOCIAL WELL-BEING PHILIPPINE PROVINCES, 1970

	Province	Education and literacy	Health	Incomes and employment	Housing and amenities
1.	Abra	0.56	-0.18	-1.24	-0.03
2.	Agusan del Norte	1.27	-0.06	0.40	0.52
3.	Agusan del Sur	-0.99	-1.68	0.41	-1.18
4.	Aklan	0.61	0.68	-1.86	-0.79
5.	Albay	-0.19	-0.39	0.29	0.07
6.	Antique	0.06	0.46	-1.31	-1.23
7.	Bataan	0.23	1.49	2.36	1.22
9.	Batanes	0.70	-0.10	-0.19	1.17
10.	Benguet	1.99	2.49	-0.04	1.74
11.	Bohol	-0.18	-0.44	0.06	0.15
12.	Bukidnon	-1.01	-1.53	0.62	-0.45
13.	Bulacan	0.61	1.56	0.96	1.95
14.	Cagayan	-0.40	-1.95	-1.45	-0.14
15.	Camarines Norte	0.41	0.60	0.07	-0.96
16.	Camarines Sur	-0.15	-0.14	-0.48	-0.17
18.	Capiz	-0.02	0.31	-0.54	-1.16
19.	Catanduanes	0.54	0.20	-0.38	-0.18
20.	Cavite	1.18	1.41	1.04	1.98
21.	Cebu	-0.04	-0.70	-0.55	-0.25
22.	Cotabato	-1.84	0.07	0.75	-0.43
23.	Cotabato, South	0.05	0.27	0.76	-0.49
24.	Davao del Norte	-0.26	-1.07	0.64	-0.38
25.	Davao del Sur	0.56	0.18	0.73	0.41
26.	Davao Oriental	-0.78	-0.88	1.32	-0.96
27.	Ifugao	-1.34	-0.60	-1.30	-0.94

Province	Education and literacy	Health	Incomes and employment	Housing and amenities
28. Ilocos Norte	1.13	0.36	-0.86	0.85
29. Ilocos Sur	0.62	0.35	-0.51	1.04
30. Iloilo	1.12	0.45	0.01	-0.28
31. Isabela	-0.94	-2.13	-0.94	-0.45
32. Kalinga-Apayao	-0.67	-3.16	0.58	-0.94
33. La Union	0.94	0.97	0.17	1.04
34. Laguna	1.16	0.82	1.07	1.71
35. Lanao del Norte	-0.04	0.83	0.93	0.69
36. Lanao del Sur	-2.07	0.24	0.39	-0.12
37. Leyte	-0.24	-1.03	-0.74	-0.36
38. Leyte, Southern	-0.23	-0.12	-0.61	-0.01
40. Marinduque	-0.07	-0.01	-0.61	0.29
41. Masbate	-1.23	-1.32	-0.94	-1.29
42. Misamis Occidental	0.92	0.54	0.04	-0.34
43. Misamis Oriental	0.87	0.11	0.43	0.49
44. Mountain Province	0.14	-0.09	0.25	0.94
45. Negros Occidental	0.43	0.11	0.71	-0.43
46. Negros Oriental	-1.10	-0.04	-0.80	-0.43
47. Nueva Ecija	0.07	0.20	-0.02	0.82
48. Nueva Vizcaya	0.19	-0.57	0.44	-0.37
49. Occidental Mindoro	-0.35	-0.51	0.81	-0.72
50. Oriental Mindoro	-0.09	-0.52	-0.46	-0.29
51. Palawan	-1.15	-0.26	-0.05	-1.50
52. Pampanga	0.23	0.85	1.60	1.86
53. Pangasinan	0.60	-1.97	-1.28	0.62
54. Quezon	0.07	-0.18	0.60	0.17
55. Rizal	3.59	1.88	3.65	3.15
56. Romblon	-0.44	0.47	-2.59	-0.89
57. Samar, Eastern	0.26	-0.29	0.16	-1.50
58. Samar, Northern	-0.87	-0.78	-0.03	-1.64
59. Samar, Western	-1.06	0.19	-0.18	-0.94
60. Sorsogon	-0.13	0.38	-1.00	-0.55
61. Sulu	-3.24	0.37	-1.14	-0.89
62. Surigao del Norte	0.77	1.10	-1.01	0.08
63. Surigao del Sur	0.35	-1.03	-0.83	-0.74
64. Tarlac	0.30	0.90	0.78	1.11
65. Zambalas	0.74	1.73	1.34	1.54
66. Zamboanga del Norte	-1.07	0.43	-0.19	-1.04
67. Zamboanga del Sur	-1.08	0.75	-0.24	-0.18

Appendix 2

FACTOR SCORES FOR PHILIPPINE PROVINCES, 1970

Province	1. Employment structure	2. Migration	3. Welfare and fertility
1. Abra	-0.54	-1.05	-0.17
2. Agusan del Norte	0.57	1.06	0.16
3. Agusan del Sur	-1.13	2.46	-0.07
4. Aklan	-0.11	-1.20	0.34
5. Albay	0.22	-0.29	0.63
6. Antique	-0.59	-1.32	0.63
7. Bataan	1.33	0.44	0.10
9. Batanes	1.08	-0.94	-0.06
10. Benguet	0.59	0.04	-2.77
11. Bohol	-0.07	-0.83	0.03

Province	1. Employment structure	2. Migration	3. Welfare and fertility
12. Bukidnon	-0.88	2.58	0.18
13. Bulacan	2.06	0.11	-0.52
14. Cagayan	-0.33	-0.46	1.26
15. Camarines Norte	0.18	0.70	0.90
16. Camarines Sur	0.12	-0.20	0.68
18. Capiz	-0.38	-0.80	0.30
19. Catanduanes	-0.14	-0.47	0.83
20. Cavite	1.83	-0.39	-0.45
21. Cebu	0.28	-0.45	-0.33
22. Cotabato	-0.77	0.95	-0.47
23. Cotabato, South	-0.28	1.63	-0.32
24. Davao del Norte	-0.45	2.27	0.24
25. Davao del Sur	0.40	1.16	-0.54
26. Davao Oriental	-0.75	2.07	0.03
27. Ifugao	-1.98	-0.42	-1.58
28. Ilocos Norte	0.19	-1.49	-0.49
29. Ilocos Sur	0.34	-1.74	-0.21
30. Iloilo	0.43	-0.82	0.05
31. Isabela	-0.40	0.53	1.12
32. Kalinga-Apayao	-1.66	1.52	-1.23
33. La Union	0.74	-0.92	-0.75
34. Laguna	1.75	0.04	-0.30
35. Lanao del Norte	0.01	0.06	-0.52
36. Lanao del Sur	-0.26	0.00	-0.00
37. Leyte	-0.46	-0.71	0.47
38. Leyte, Southern	-0.19	-0.84	1.48
40. Marinduque	0.09	-0.76	0.40
41. Masbate	-1.01	-0.04	0.70
42. Misamis Occidental	0.17	-0.17	0.32
43. Misamis Oriental	0.24	0.28	-0.11
44. Mountain Province	-1.84	-1.20	-5.43
45. Negros Occidental	0.00	-0.43	-0.38
46. Negros Oriental	-0.62	-0.80	0.35
47. Nueva Ecija	1.14	-0.11	0.99
48. Nueva Vizcaya	-0.35	0.72	0.53
49. Occidental Mindoro	-0.02	1.74	1.00
50. Oriental Mindoro	-0.28	0.22	0.52
51. Palawan	-0.98	1.20	0.30
52. Pampanga	1.90	0.21	0.07
53. Pangasinan	0.78	-0.89	0.47
54. Quezon	0.10	0.17	0.11
55. Rizal	4.00	1.50	-1.51
56. Romblon	-0.17	-1.41	1.32
57. Samar, Eastern	-0.77	-0.34	0.61
58. Samar, Northern	-1.02	-0.09	1.22
59. Samar, Western	-0.88	-0.55	0.36
60. Sorsogon	-0.13	-0.33	1.06
61. Sulu	-1.02	-1.26	-0.18
62. Surigao del Norte	-0.41	-0.86	-0.13
63. Surigao del Sur	-0.65	0.56	-0.02
64. Tarlac	1.00	-0.36	0.15
65. Zambales	1.65	0.68	-0.84
66. Zamboanga del Norte	-1.09	-0.05	-0.12
67. Zamboanga del Sur	-0.58	0.10	-0.41

Appendix 3

FACTOR SCORES FOR MUNICIPALITIES OF REGION 6 (WEST VISAYAS), 1970

Province	Municipality	(1) Urban employment	(2) Population structure
Aklan	01) Kalibo	2.96	0.78
	02) Altaras	-0.35	-0.69
	03) Balete	-0.88	0.34
	04) Banga	0.97	-1.44
	05) Batan	-0.25	-1.30
	06) Buruanga	-0.70	-0.42
	07) Ibajay	0.05	-1.12
	08) Lezo	1.87	-2.40
	09) Libacao	-1.92	0.86
	10) Madalag	-1.73	-0.47
	11) Makato	0.31	-0.90
	12) Malay	0.23	-1.21
	13) Malinao	-0.59	-0.91
	14) Nabas	0.66	-0.92
	15) New Washington	-0.11	-0.26
	16) Numancia	1.83	-0.74
	17) Tangalan	-0.33	-0.17
Antique	01) San Jose	2.40	-0.78
	02) Anini-y	0.49	-0.57
	03) Barbaza	-1.06	-0.05
	04) Belison	0.81	-1.21
	05) Bugasong	-0.33	-0.65
	06) Caluya	-0.93	-0.11
	07) Culasi	-0.12	-1.29
	08) Dao	-0.03	-0.49
	09) Hamtic	0.08	-0.10
	10) Lawa-an	-1.22	-0.36
	11) Libertad	-0.49	-0.14
	12) Pandan	-0.31	-0.91
	13) Patnongon	-0.66	-0.53
	14) San Remigio	-1.67	0.14
	15) Sebaste	-0.94	-1.66
	16) Sibalom	0.19	-1.01
	17) Tibiao	-0.21	-1.14
	18) Valderrama	-1.43	-0.35
Capiz	01) Roxas City	1.87	0.39
	02) Cuartero	-0.35	0.53
	03) Dao	-0.82	-0.11
	04) Dumalag	0.18	-0.53
	05) Dumarao	-0.88	-0.07
	06) Ivisan	0.26	-0.37
	07) Jamindan	-1.66	0.36
	08) Ma-Ayon	-1.39	0.90
	09) Mambusao	0.14	-0.07
	10) Panay	-0.65	-0.14
	11) Panitan	-0.75	0.05
	12) Pilar	-0.55	-0.67
	13) Pontevedra	-0.38	0.25
	14) President Roxas	1.05	0.72
	15) Sapi-An	-0.01	-0.24
	16) Sigma	-1.11	-0.46
	17) Tapaz	-1.68	1.44

Province	Municipality	(1) Urban employment	(2) Population structure
Iloilo	01) Iloilo City	4.05	1.78
	02) Ajuy	-0.56	-0.95
	03) Alimodian	0.05	-0.41
	04) Anilao	-0.85	0.49
	05) Badiangan	0.14	-1.58
	06) Balasan	-0.24	1.12
	07) Banate	0.17	-0.55
	08) Barotac Nuevo	0.73	-0.24
	09) Barotac Viejo	0.52	-0.52
	10) Batad	-1.10	0.41
	11) Bingawan	-0.62	0.81
	12) Buenavista	-0.12	-0.48
	13) Cabatuan	0.98	-1.37
	14) Calinog	-0.36	0.61
	15) Carles	-0.95	-0.18
	16) Concepcion	-0.74	0.12
	17) Dingle	0.24	-0.37
	18) Duenas	-0.55	-0.39
	19) Dumangas	0.43	-0.64
	20) Estancia	-0.37	0.25
	21) Guimbal	1.18	-1.34
	22) Igbaras	-0.24	-1.01
	23) Janiuay	0.35	0.09
	24) Jordan	-0.46	-0.02
	25) Lambunao	-0.52	-0.07
	26) Leganes	0.77	-0.39
	27) Lemery	-0.99	0.66
	28) Leon	0.05	-0.06
	29) Maasin	-0.10	-0.47
	30) Miagao	0.83	-1.13
	31) Mina	0.27	-1.01
	32) New Lucena	0.46	-3.13
	33) Nueva Valencia	0.04	0.72
	34) Oton	1.50	-0.98
	35) Passi	0.16	0.45
	36) Pavia	0.45	-0.93
	37) Pototan	1.20	-0.97
	38) San Dionisio	-0.77	-0.10
	39) San Enrique	0.14	-0.41
	40) San Joaquin	-0.21	-0.82
	41) San Miguel	0.59	-0.59
	42) San Rafael	-1.05	0.41
	43) Santa Barbara	1.82	-0.80
	44) Sara	-0.26	-0.09
	45) Tigbauan	0.44	-1.05
	46) Tubungan	-0.92	-1.02
	47) Zarraga	0.06	-0.47
Negros Occidental	01) Bacolod City	3.96	2.59
	02) Bago City	0.21	1.12
	03) Binalbagan	0.70	1.04
	04) Cadiz City	0.19	2.48
	05) Calatrava	-1.11	1.10
	06) Candoni	-1.30	1.31
	07) Cauayan	-0.97	0.66
	08) Enrique B Magalo	0.00	1.39
	09) Escalante	-0.65	0.85

Province	Municipality	(1) Urban employment	(2) Population structure
Negros Occidental (contd.)	10) Himamaylan	-0.20	0.87
	11) Hinigaran	0.47	0.46
	12) Hinoba-An	-0.53	2.27
	13) Ilog	0.03	0.08
	14) Isabela	-0.13	0.54
	15) Kabankalan	-0.74	1.27
	16) La Carlota City	1.80	1.31
	17) La Castellana	-0.16	0.73
	18) Manapla	0.05	2.08
	19) Moises Padilla	-1.00	0.23
	20) Murcia	-0.49	1.19
	21) Pontevedra	0.23	0.28
	22) Pulupandan	1.46	0.84
	23) Sagay	0.16	1.47
	24) San Carlos City	-0.51	1.44
	25) San Enrique	0.73	0.57
	26) Silay City	0.97	2.00
	27) Sipalay	0.08	2.21
	28) Talisay	0.96	0.83
	29) Toboso	-0.73	0.97
	30) Valladolid	0.49	0.22
	31) Victorias	1.43	2.46

Appendix 4

SELECTED REFERENCES AND APPLICATIONS OF PRINCIPAL COMPONENTS ANALYSIS AND FACTOR ANALYSIS

Berri, B.J.L. and Ray, D.M. (1966), "Multivariate socio-economic regionalization: A pilot study in central Canada" in Rymes T. and Ostry S. (eds.) Regional Statistical Studies, University of Toronto Press, Toronto.

Blaikie, P.M. (1971), "Spatial organization of agriculture in some north Indian villages", Transactions of the Institute of British Geographers No. 52, 1-40.

Cattell, R.B. (1965), "Factor analysis: An introduction to essentials, I and II", Biometrics Vol. 21, 190-215 and 405-435.

Clark, D., Davies, W.K.D. and Johnston, R.J., "The application of factor analysis in human geography", The Statistician, Vol. 23, No. 3/4, 259-281.

Cooley, W.W. and Lohnes, P.R. (1971), Multivariate data analysis, John Wiley and Sons, New York.

Greer-Wootten, B. (1972), A bibliography of statistical applications in geography, Association of American Geographers, Commission on College Geography, Technical Paper No. 9, Washington D.C.

Guertin, W.H. and Bailey, J.P. (1970), Introduction to modern factor analysis, Edwards Bros. Ann Arbor, Michigan.

Harman, H.H. (1966), Modern factor analysis, second edition, Chicago University Press, Chicago.

Hartley, R.G. and Norris, J.M. (1969), "Demographic regions of Libya: A principal components analysis of economic and demographic variables", Tijdschrift voor Economische en sociale geografie, Vol. 60, 221-227.

Horst, P. (1965), Factor analysis of data matrices, Holt Rinehard & Winston, New York.

King, L.J. (1969), Statistical analysis in geography, Prentice Hall, Englewood Cliffs.

Lawley, D.N. and Maxwell, A.E. (1963), Factor analysis as a statistical method, Butterworth and Co., London.

Mabogunje, A.L. (1965), "Urbanization in Nigeria: A constraint on economic development", Economic development and cultural change, Vol. 13, 413-438.

Rees, P.H. (1971), "Factorial ecology: An extended definition, survey and critique of the field", Economic geography, Vol. 47, 220-233.

Rummel, R.J. (1970), Applied factor analysis, Northwestern University Press, Chicago.

PHILIPPINE PROVINCES

1. Abra
2. Agusan del Norte
3. Agusan del Sur
4. Aklan
5. Albay
6. Antique
7. Bataan
8. Batanes
9. Batangas
10. Benguet
11. Bohol
12. Bukidnon
13. Bulacan
14. Cagayan
15. Camarines Norte
16. Camarines Sur
17. Camiguin
18. Capiz
19. Catanduanes
20. Cavite
21. Cebu
22. Cotabato
23. Cotabato, South
24. Davao del Norte
25. Davao del Sur
26. Davao Oriental
27. Ifugao
28. Ilocos Norte
29. Ilocos Sur
30. Iloilo
31. Isabela
32. Kalinga-Apayao
33. La Union
34. Laguna
35. Lanao del Norte
36. Lanao del Sur
37. Leyte
38. Leyte, Southern
39. Manila
40. Marinduque
41. Masbate
42. Misamis Occidental
43. Misamis Oriental
44. Mountain Province
45. Negros Occidental
46. Negros Oriental
47. Nueva Ecija
48. Nueva Vizcaya
49. Occidental Mindoro
50. Oriental Mindoro
51. Palawan
52. Pampanga
53. Pangasinan
54. Quezon
55. Rizal
56. Romblon
57. Samar, Eastern
58. Samar, Northern
59. Samar, Western
60. Sorsogon
61. Sulu
62. Surigao del Norte
63. Surigao del Sur
64. Tarlac
65. Zambales
66. Zamboanga del Norte
67. Zamboanga del Sur

Appendix

List of Documents

A. Human resources indicators

1. Progress Report I on the Methodology of Human Resources Indicators. (UNESCO/COM/SHC/CS/194/5)

2. Study I. Pilot study for the development of an estimating formula for the total stock of high-level manpower: Manpower components approach.

3. Study II. An analytical approach to educational attainment indicators based on the analysis of school enrolment by grade.

4. Study III. Procedure of evaluating high-level manpower data and typology of countries by means of taxonomic method, by Z. Hellwig.

5. Study IV. On the definition and theory of development with a view to the application of rank order indicators in the elaboration of a composite index of human resources. Paper prepared for the Unesco Expert Meeting on the Methodology of Human Resources indicators, held in Warsaw, 11-16 December 1967, by Johan Galtung and Tord Hoivik. (The paper can also be identified as PRIO-publication No. 6-4 from the International Peace Research Institute, Oslo.)

6. Study V. Discussion of methodological guidelines, problem areas and difficulties connected with the elaboration of coherent systems of indices of human resources, by Z. Gostkowski.

7. Study VI. On the optimal choice of predictors, by Z. Hellwig.

8. Study VII. On the problem of weighting in international comparisons, by Z. Hellwig.

9. Study VIII. Indices of development for selected Latin American and middle African countries. An experimental exercise using the Hellwig taxonomic method, by Frederick H. Harbison, Joan Maruhnic, Jane R. Resnick.

10. Study IX. Alternative combinations of human resources components as predictors of economic growth in Latin America, by the Higher School of Economics, Wroclaw, Poland.

11. Study X. Concerning data availability and conceptual validity. Selected Latin American and African countries, by Frederick H. Harbison, Joan Maruhnic and Jane R. Resnick.

12. Study XI. The assessment of the validity of human resources indicators by means of a Cobb-Douglas type production function, by H. V. Muhsam, Hebrew University, Jerusalem, Israel.

13. Study XII. The use of taxonomic measures in target setting based on international comparisons, by Z. Gostkowski.

14. Study XIII. Human resources and economic growth: The case of Mexico, by Tord Hoivik, International Peace Research Institute, Oslo.

15. Study XIV. Human resources and socio-economic development:
I. Theory, methods and data; II. The case of Japan, by Johan Galtung.

16. Study XV. Human resources and socio-economic development: The case of Venezuela, by Kristin Tornes, International Peace Research Institute, Oslo.

17. Study XVI. Approximate methods of selection of an optimal set of predictors, by Z. Hellwig.

18. Study XVII. Mathematical methods in the Unesco Project on Human Resources Indicators: (i) An evaluation, (ii) Alternative analytical methods, by Ludovic Lebart (SHC/WS/219).

19. Study XVIII. Human resources and economic development. Some problems on measurements, by Professor Wilfred Beckerman (University College, London).

20. Study XIX. Distance-based analysis, numerical taxonomy and classification of countries according to selected areas of socio-economic development, by S. Fanchette, Methods and Analysis Division, Unesco.

21. Study XX. Diachronic analysis of relationships between human resources components and the rate of economic growth in selected countries by Johan Galtung. (COM/WS/131)

22. Study XXI. The selection of a set of "core" indicators of socio-economic development, by Z. Hellwig and S. Fanchette.

23. Study XXII. The quest of an employment strategy in developing countries and its relationship to the work on human resources indicators, by H. W. Singer (Institute of Development Studies, The University of Sussex, United Kingdom.)

24. Paper "Synchronic and diachronic approaches in the Unesco project on human resources indicators. Wroclaw taxonomy and bivariate diachronic analysis", by Serge Fanchette - Methods and Analysis Division, Unesco. (SHC/WS/209)

25. Study XXIV. A method of establishing a list of development indicators, by B. Ivanovic, June 1973. (In French and in English)

26. Distance-based analysis, numerical taxonomy and classification of countries according to selected areas of socio-economic development, May 1972. (SHC/WS/237)

27. Quantitative analysis of modernization and development by Frederick H. Harbison, Joan Maruhnic, Jane R. Resnick.

B. **Indicators of economic and social change and their use in development planning**

28. Study XXV. Toward a system of quantitative indicators of components of human resources development, by B. Ivanovic and S. Fanchette.

29. Study XXVI. Aspects of the distribution of income and wealth in Kenya, by H. W. Singer and Stuart D. Reynolds.

30. Study XXVII. Distributional patterns of Development and Welfare: A case study of Zambia, by Charles Eliot.

31. Study XXVIII. Typological study using the Wrocklaw taxonomic method (A study of regional disparities in Venezuela), by José F. Silvio-Pomenta.

32. Report of informal consultations on indicators of economic and social change, held at Unesco Headquarters, 22-25 October 1973. (SHC/683/1403)

33. "Theories, models and indicators of social change", by Kenneth C. Land (Department of Sociology, University of Illinois at Urbana Champaign). (SHC/74/WS/1, 9 April 1974)

34. "Esquisse méthodologique de la modélisation sociale", par Gérard Martin (Docteur ès sciences économiques, Maître assistant à l'Institut d'études politiques de l'Université de Grenoble). (SHC/74/WS/6, 3 May 1974)

35. Report of regional consultations on indicators of economic and social change, held at Unesco Headquarters, 20-22 May 1974. (SHC/499/020376)

36. "On the construction of social indicators" (illustrated with Indian examples) by Ramkrishna Mukherjee (Indian Statistical Institute, Calcutta). (SHC/74/WS/20, 2 August 1974)

37. "L'expérience française en planification sociale bilan et perspectives", par Gérard Martin (Institut d'études politiques de l'Université de Grenoble). (SHC/74/WS/23, 14 August 1974)

38. "The spatial dimensions of national planning: the rôle of territorial socio-economic indicators in the formulation and implementation of national development plans in the Asian region", by R. G. Cant (Department of Geography, University of Canterbury, New Zealand and Research School of Pacific Studies Australian National University). (SHC/WS/24, 1 September 1974)

39. "Socio-economic indicators for development planning", by M. V. S. Rao (Joint Director, Central Statistical Organization, Government of India). (SHC/74/WS/21, 1 September 1974)

40. "Some observations on the use of social indicators in development planning, with special reference to levels of livings", by Stephen H. K. Yeh, (Professor of Sociology and Urban and Regional Planning, University of Hawaii). (SHC/74/WS/26, 2 September 1974)

41. Complete set of H. R. I. P. papers including panel meetings.

42. Report meeting of a working group on the application of socio-economic indicators to development planning, Bangkok, 16-24 September 1974. (SHC/75/WS/4)

43. Workshop on the application of social indicators to national planning in Thailand, Bangkok, 23 January 1976. Special patterns and regional structures in Thailand: An application of territorial indicators as an input in the development planning process, by R. G. Cant. (SHC/75/WS/57)

44. Workshop on the application of social indicators to national planning in the Philippines, Manila, 26 January, Iloilo, 28-29 January. Planning for development in the Philippines: case studies to examine the applicability of territorial indicators at national and regional levels, by R. G. Cant. (SHC/75/WS/58)

45. Social indicators for development planning. A Philippine case study, Mahar Mangahas. (SHC/75/WS/60)

46. Toward a system of human resources indicators for less developed countries. A selection of papers prepared for Unesco research project, edited by Zygmunt Gostkowski.

47. Socio-economic indicators theories and applications. International social science journal, Vol. XXVII, No. 1, 1975, Unesco.

48. The use of socio-economic indicators in development planning, edited by Nancy Baster. The Unesco Press, 1976, ISBN 92-3-101329-7.

49. Social indicators: problems of definition and of selection. Report and papers in the social sciences, No. 30, Unesco.

50. The integration of social indicators in planning process, by Gérard Martin. (SHC/75/WS/52, 16 January 1976)

51. The place of internal migration and urbanization in the frame of a system of indicators of socio-economic development, by H. V. Muhsam.

(Hebrew University, Jerusalem, Israel). (SHC/75/WS/54, 23 January 1976)

52. Human needs, human rights and the theories of development, by Johan Galtung and Anders Wirak (Chair in conflict and peace research, University of Oslo, Institut d'Etudes du Développement, Geneva). (SHC/75/WS/55, 20 January 1976)

53. Progress report of Unesco project on indicators of economic and social change and their applicability in development planning. Secretariat background paper. (SHC/76/WS/11, 8 March 1976)

54. Report of informal consultations on indicators of economic and social change, held at Unesco Headquarters, 20-22 May 1974.

55. Vers un système d'indicateurs correspondant au modèle de la stratégie internationale du développement, par Branislav Ivanovic. (SHC/76/WS/14, March 1976)

56. Social ends and social indicators in national economy planning, by G. V. Osipov. (SS. 76/WS/21, April 1976)

57. Some current work on indicators of social and economic change in the U.S.A., by Frank M. Andrews. (SS. 76/WS/22, April 1976)

58. The use of indicators in development planning in the Sudan, by Nancy Baster. (SS. 76/WS/24, April 1976)

Argentina	EDILYR, Belgrano 2786-88 BUENOS-AIRES.
Australia	*Publications*: Educational Supplies Pty. Ltd., Box 33, Post Office, BROOKVALE 2100, N.S.W. *Periodicals*: Dominie Pty. Ltd., Box 33, Post Office, BROOKVALE 2100, N.S.W. *Sub-Agent*: United Nations Association of Australia (Victorian Division), 5th Floor, 134–136 Flinders St., MELBOURNE 3000.
Austria	Dr Franz Hain, Verlags- und Kommissionsbuchhandlung, Industriehof Stadlau, Dr Otto-Neurath-Gasse 5, 1220 WIEN.
Belgium	Jena De Lannoy, 112, rue du Trône, 1050 BRUXELLES. CCP 000-0070823-13.
Bolivia	Los Amigos del Libro: casilla postal 4415, LA PAZ; Peru 3712 (Esq. Espana), casilla postal 450, COCHABAMBA.
Brazil	Fundação Getúlio Vargas, Serviço de Publicações, caixa postal 21120 Praia de Botafogo 183, RIO DE JANEIRO G.B.
Bulgaria	Hemus, Kantora Literatura, bd. Rousky 6, SOFIJA.
Burma	Trade Corporation n.º (9), 550–552 Merchant Street, RANGOON.
Canada	Renouf Publishing Company Ltd., 2182 St. Catherine Street West, MONTREAL, Que. H3H 1M7.
Chile	Bibliocentro Ltda., casilla 13731, Huérfanos 1160 of. 213, SANTIAGO (21).
Colombia	Editorial Losada Ltda., calle 18A, n.ºs 7-37, apartado aéreo 5829, apartado nacional 931. BOGOTÁ. J. Germán Rodríguez N., Calle 17, 6–59, apartado nacional 83, GIRARDOT (Cundinamarca). Librería Buccholz Galería, avenida Jiménez de Quesada 8–40, apartado aéreo 53750, BOGOTÁ. *Subdepots*: Edificio La Ceiba, Oficina 804. MEDELLÍN; Calle 37, n.ºs 14–73, Oficina 305, BUCARAMANGA; Edificio Zaccour, Oficina 736, CALI.
Congo	Librairie populaire, B.P. 577, BRAZZAVILLE.
Costa Rica	Librería Trejos S.A., apartado 1313, SAN JOSÉ. Teléfonos: 2285 y 3200.
Cuba	Instituto Cubano del Libro, Centro de Importación, Obispo 461, LA HABANA.
Cyprus	'MAM', Archbishop Makarios 3rd Avenue, P.O. Box 1722, NICOSIA.
Czechoslovakia	SNTL, Spalena 51, PRAHA 1 (*Permanent display*); Zahranicni literatura, 11 Soukenicka, PRAHA 1. *For Slovakia only*: Alfa Verlag, Publishers, Hurbanovo nam. 6. 803 31 BRATISLAVA.
Dahomey	Librairie nationale, B.P. 294, PORTO NOVO.
Denmark	Ejnar Munksgaard Ltd., 6 Nørregade, 1165 KØBENHAVN K.
Egypt	National Centre for Unesco Publications, 1 Talaat Harb Street, Tahrir Square, CAIRO.
El Salvador	Librería Cultural Salvadoreña, S.A., calle Delgado, n.º 117, SAN SALVADOR.
Ethiopia	Ethiopian National Agency for Unesco, P.O. Box 2996, ADDIS ABABA.
Finland	Akateeminen Kirjakauppa, 2 Keskuskatu, SF-00100 HELSINKI 10.
France	Librairie de l'Unesco, 7 place de Fontenoy, 75700 PARIS. CCP Paris 12598-48.
French West Indies	Librairie 'Au Boul' Mich', 1 Rue Perrinon *and* 66 Avenue du Parquet, 97200 FORT-DE-FRANCE (Martinique).
German Dem. Rep.	International bookshops *or* Buchhaus Leipzig, Postfach 140, 701 LEIPZIG.
Fed. Rep. of Germany	Verlag Dokumentation, Pössenbacherstrasse 2, 8000 MÜNCHEN 71 (Prinz Ludwigshöhe). '*The Courier*' (German *edition only*): Colmantstrasse 22, 5300 BONN. *For scientific maps only*: GEO Center, Postfach 800830, 7000 STUTTGART 80.
Ghana	Presbyterian Bookshop Depot Ltd., P.O. Box 195, ACCRA; Ghana Book Suppliers Ltd., P.O. Box 7869, ACCRA; The University Bookshop of Ghana, ACCRA; The University Bookshop of Cape Coast; The University Bookshop of Legon, P.O. Box 1, LEGON.
Greece	Main bookshops in Athens (Eleftheroudakis, Kauffman, etc.).
Hong Kong	Federal Publications Division, Far East Publication Ltd., 5 A Evergreen Industrial Mansion, Wong Chuk Hang Road, ABERDEEN; Swindon Book Co., 13–15 Lock Road, KOWLOON.
Hungary	Akadémiai Könyvesbolt Váci y 22, BUDAPEST V. A.K.V. Könyvtárosok Boltja, Népköztársaság utja 16. BUDAPEST VI.
Iceland	Snaebjörn Jonsson & Co. H. F., Hafnarstracti 9, REYKJAVIK.
India	Orient Longman Ltd.: Kamani Marg, Ballard Estate, BOMBAY 400038; 17 Chittaranjan Avenue, CALCUTTA 13; 36A Anna Salai, Mount Road, MADRAS 2; B-3/7 Asaf Ali Road, NEW DELHI 1; 80/1 Mahatma Gandhi Road, BANGALORE 560001; 3-5-820 Hyderguda, HYDERABAD 500001. *Sub-depots*: Oxford Book and Stationery Co., 17 Park Street, CALCUTTA 700016 *and* Scindia House, NEW DELHI 110001; Publication Section, Ministry of Education and Social Welfare, 511 C-Wing, Shastri Bhavan, NEW DELHI 110001.
Indonesia	Bhratara Publishers and Booksellers, 29 Jl. Oto Iskandardinata III, JAKARTA. Gramedia Bookshop, Jl. Gadjah Mada 109, JAKARTA. Indira P.T., Jl. Dr. Sam Ratulangie, 37 JAKARTA PUSAT.
Iran	Commission nationale iranienne pour l'Unesco, avenue Iranchahr Chomali n° 300, B.P. 1533, TÉHÉRAN. Kharazmie Publishing and Distribution Co. 220 Daneshgahe Street, Shah Avenue, P.O. Box 14/1486, TÉHÉRAN.
Iraq	McKenzie's Bookshop, Al-Rashid Street, BAGHDAD.
Ireland	The Educational Company of Ireland Ltd., Ballymount Road, Walkinstown, DUBLIN 12.
Israel	Emanuel Brown, formerly Blumstein's Bookstores: 35 Allenby Road *and* 48 Nachlat Benjamin Street, TEL AVIV; 9 Shlomzion Hamalka Street, JERUSALEM.
Italy	LICOSA (Liberia Commissionaria Sansoni S.p.A.), via Lamarmora 45, casella postale 552, 50121 FIRENZE.
Jamaica	Sangster's Book Stores Ltd., P.O. Box 366, 101 Water Lane, KINGSTON.
Japan	Eastern Book Service Inc., C.P.O. Box 1728, TOKYO 100 92.
Kenya	East African Publishing House, P.O. Box 30571, NAIROBI.
Republic of Korea	Korean National Commission for Unesco, P.O. Box Central 64, SEOUL.
Kuwait	The Kuwait Bookshop Co. Ltd., P.O. Box 2942, KUWAIT.
Lesotha	Mazenod Book Centre, P.O. MAZENOD.
Liberia	Cole & Yancy Bookshops Ltd., P.O. Box 286, MONROVIA.
Libya	Agency for Development of Publication and Distribution, P.O. Box 34–35, TRIPOLI.
Luxembourg	Librairie Paul Bruck, 22 Grande-Rue, LUXEMBOURG.
Madagascar	Commission Nationale de la République démocratique de Madagascar pour l'Unesco, B.P. 331, TANANARIVE.
Malaysia	Federal Publications Sdn. Bhd., Balai Berita, 31 Jalan Riong, KUALA LUMPUR.
Malta	Sapienzas, 26 Republic Street, VALLETTA.
Mauritius	Nalanda Co. Ltd., 30 Bourbon Street, PORT-LOUIS.
Mexico	*For publications only*: CILA (Centro Interamericano de Libros Académicos), Sullivan 31 *bis*, MÉXICO 4 D.F. *For publications and periodicals*: SABSA, Servicios a Bibliotecas, S.A., Insurgentes Sur n.º 1032-401, MÉXICO 12 D.F.
Monaco	British Library, 30, boulevard des Moulins, MONTE-CARLO.
Mozambique	Instituto Nacional do Livro e do Disco (INLD), avenida 24 de Julho 1921, r/c e 1º andar, MAPUTO.
Netherlands	N.V. Martinus Nijhoff, Lange Voorhout 9, 's-GRAVENHAGE; Systemen Keesing, Ruysdaelstraat 71-75, AMSTERDAM 1007.
Netherlands Antilles	G. C. T. Van Dorp & Co. (Ned. Ant.) N.V., WILLEMSTAD (Curaçao, N.A.).
New Caledonia	Reprex S.A.R.L., B.P. 1572, NOUMÉA.
New Zealand	Government Printing Office, Government Bookshops: Rutland Street P.O. Box 5344, AUCKLAND; 130 Oxford Terrace, P.O. Box 1721, CHRISTCHURCH; Alma Street, P.O. Box 857, HAMILTON; Princes Street, P.O. Box 1104, DUNEDIN; Mulgrave Street, Private Bag, WELLINGTON.
Niger	Librairie Manclert, B.P. 868, NIAMEY.
Nigeria	The University Bookshop of Ife; The University Bookshop of Ibadan, P.O. Box 286, IBADAN; The University Bookshop, Nsukka; The University Bookshop of Lagos; The Ahmadu Bello University Bookshop of Zaria.
Norway	*All publications*: Johan Grundt Tanum, Karl Johans gate 41/43, OSLO 1. '*The Courier*' *only*: A/S Narvesens Litteraturjeneste, Box 6125, OSLO 6.
Pakistan	Mirza Book Agency, 65 Shahrah Quaid-e-azam, P.O. Box 729, LAHORE 3.
Peru	Editorial Losada Peruana, Jirón Contumaza 1050, apartado 472, LIMA.
Philippines	The Modern Book Co., 926 Rizal Avenue, P.O. Box 632, MANILA D-404.
Poland	Ars Polona-Ruch, Krakowskie Przedmiescie n.º 7, 00-901 WARSZAWA. ORPAN-Import, Palac Kultury, 00-901 WARSZAWA.
Portugal	Dias & Andrale Ltda., Libraria Portugal, rua o Carmo 70, LISBOA.
Southern Rhodesia	Textbook Sales (PVT) Ltd., 67 Union Avenue, SALISBURY.
Romania	I.C.E. LIBRI, Calea Victoriei nr. 126, P.O. Box 134-135 BUCUREŞTI. *Subscriptions to periodicals*: Rompresfilatelia, Calea Victoriei nr. 29, BUCUREŞTI.